MATLAB® Handbook with Applications to Mathematics, Science, Engineering, and Finance

José Miguel David Báez-López

David Alfredo Báez Villegas

CRC Press
Taylor & Francis Group
Boca Raton London New York

CRC Press is an imprint of the
Taylor & Francis Group, an **informa** business

A CHAPMAN & HALL BOOK

CRC Press
Taylor & Francis Group
6000 Broken Sound Parkway NW, Suite 300
Boca Raton, FL 33487-2742

First issued in paperback 2020

ISBN-13: 978-1-138-62645-4 (hbk)
ISBN-13: 978-0-367-73203-5 (pbk)

Library of Congress Cataloging-in-Publication Data

Names: Báez López, David, author. | Báez Villegas, David Alfredo, author.
Title: MATLAB handbook with applications to mathematics, science, engineering,
  and finance / José Miguel David Báez-López and David Alfredo Báez Villegas.
Description: Boca Raton, Florida : CRC Press, [2019].
Identifiers: LCCN 2018030135| ISBN 9781138626454 (hardback : alk. paper) |
  ISBN 9781315228457 (ebook).
Subjects: LCSH: Numerical analysis--Data processing. | MATLAB.
Classification: LCC QA297 .B27 2019 | DDC 510.285/536--dc23
LC record available at https://lccn.loc.gov/2018030135

Visit the Taylor & Francis Web site at
http://www.taylorandfrancis.com

and the CRC Press Web site at
http://www.crcpress.com

To Gary and Laura, my children and Ofelia, my wife.
David Báez-López

Contents

Preface

Mathematics is used in almost every field of knowledge. It is a necessary tool in engineering, physics, science, finance, biology, chemistry, and accounting, to name a few. Every day, new disciplines emerge that require massive computational effort. We can mention Artificial Intelligence as an example of them. Mathematics is taught at all levels of education, from kindergarten and elementary school to college and graduate school. Thus, most people have a fairly good knowledge of some area of mathematics. Unfortunately, most students and mathematics users are not taught using mathematics software tools like MATLAB®[1], among other similar tools such as Mathematica and MathCAD, which allow users to solve mathematics problems when they arise in their corresponding fields of expertise. The purpose of this handbook is to allow mathematics users to learn and master the mathematics software package MATLAB.

MATLAB integrates computation, visualization, and programming to produce a powerful tool for a number of different tasks in mathematics. MATLAB is the acronym for MATrix LABoratory. In MATLAB, every mathematical variable or quantity is treated in matrix form.

With MATLAB, we can perform complex mathematical tasks with relatively simple programs. This is possible because MATLAB has close to 10,0000 built-in functions, from simple ones such as differentiation, integration, and plotting, to optimization functions which require no user programming.

Many of the functions mentioned above are grouped into toolboxes specially dedicated to some field of science, finance, or engineering.

Another important topic covered in the handbook is object-oriented programming, the paradigm that allows us to create graphical user interfaces. The main concepts of this paradigm, as well some examples, are presented.

Simulink® is another software package that runs from MATLAB. Simulink simulates systems at the block level. Thus, it is ideal for scientific and engineering system simulation. Simulink is described in Chapter 9 and some examples show the great advantages of using it for system modelling and simulation.

There are many books available on MATLAB, but a unique feature of this handbook is that it can be used by novices and experienced users alike. It is written for the first time MATLAB user who wants to learn the basics of MATLAB, but, at the same time, it can be used by users with a basic MATLAB

---

[1]For contact information write to Mathworks, Inc., 3 Apple Hill Dr., Natick, MA 01760-2098, USA, E-mail: info@mathworks.com, Web: www.mathworks.com.

knowledge who want to learn advanced topics such as programming, creating executables, publishing results directly from MATLAB programs, and creating graphical user interfaces. In addition, for experienced users, it has chapters with MATLAB applications in engineering, physics, finance, image processing, and optimization. Each and every one of the examples and exercises were solved using MATLAB Releases 2017b and 2018b.

The authors wish to thank those in the Book program from The MathWorks, Inc. The authors also wish to thank the staff at Taylor & Francis Group.

David Báez-López
David Alfredo Báez Villegas

List of Figures

List of Tables

Chapter 1

Introduction to MATLAB

1.1 Introduction

MATLAB® is a very high-level powerful system designed for technical computing. It integrates in the same software environment computation, programming, and visualization. It also has a very easy mathematical notation. Some of the most common applications for technical computing are as varied as:

Algorithmic development
Modeling and simulation
Data analysis
Plotting
Graphical User Interfaces
Rapid prototyping
Toolboxes
Vectorization

MATLAB is an acronym for **MAT**rix **LAB**oratory and it was originally developed to perform matrix calculations. MATLAB includes a programming language which is, probably, more powerful than traditional programming languages such as C, C++, C#, VisualBasic, and Python, to name a few.

MATLAB was developed in 1984 by Cleve Moler and Jack Little, who founded The MathWorks, Inc. in Natick, Massachusetts. The first version of MATLAB only had about eighty functions. The very latest version includes more than ten thousand functions.

Besides MATLAB, The MathWorks has developed a series of software packages, called toolboxes, written in the MATLAB programming language. These toolboxes can perform a number of calculations in several branches of engineering, economics, finance, physics, and mathematics, among others. It is hard to imagine an area of knowledge where MATLAB does not have an application. In the coming chapters we will use examples of the different areas where MATLAB has applications to illustrate how to use MATLAB.

1.1.1   Book Organization

The book is organized in the following way. The first two chapters are an introduction to MATLAB. The following three chapters, 3 to 5, cover basic calculations in MATLAB. The topics include linear algebra, calculus, and plotting. The next three chapters, 6 to 9, cover programming, advanced programming techniques, graphical user interface (GUI) development, and Simulink® which is a MATLAB-based GUI useful for system modeling and simulation. The last five chapters cover a broad set of examples illustrating applications in several engineering disciplines, physics, and finance.

1.1.2   Chapter Organization

The chapter is organized as follows. It begins with a MATLAB environment description. It describes the basic layout desktop and explains the purpose of the different windows available. It continues with basic calculations describing how to create and store variables. Different formats for variables are introduced. Basic functions are presented and the set of elementary function is described. Strings are introduced and some operations on strings are described. A very useful characteristic of MATLAB is the fact that all the variables, functions, and calculations are stored in the `Command History` and thus, they can be stored in a file, or a program can be created from them. This is covered in detail in this chapter. Finally, it is explained how to use MATLAB Help.

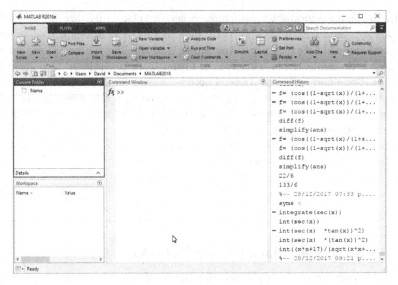

FIGURE 1.1: MATLAB main window.

1.2   Starting MATLAB

To install MATLAB, follow the instructions provided by The MathWorks, Inc. to install the software and the licenses. When the installation is finished, a MATLAB icon will be displayed on the Desktop and in the Programs path. When we click on the icon, MATLAB will open, as shown in Figure 1.1. In the main menu in the Desktop we can display up to four windows:

1. `Command Window`,

2. `Workspace`,

3. `Current Directory`, and

4. `Command History`.

Each of these windows can be attached to the main window, but we can have any of them as a separate window by using the right upper arrow. We can reattach the window using the same arrow.

The `Command window` is the window where we enter variable values and instructions to perform calculations. The `Workspace window` contains the variables created in the current session. The `Current Directory` displays the files and folders in the current directory. The `Command History` displays the instructions executed in the current and previous sessions, unless it has been deleted. Each of these windows has a pulldown menu that can be displayed with the triangle at the right-hand upper corner for each window.

In the main window we also have a set of tabs which are HOME, PLOTS, and APPS. The HOME tab is shown by default and it contains a toolbar with standard icons like New Script, New, Open, and the like. There are some icons to Import Data and Save Workspace. Other icons are related to coding like Analyze Code and Run and Time. There is an icon for Simulink, the systems simulator which is covered in Chapter 9. Several of the icons in the HOME toolbar are discussed and used in the book. The PLOTS tab shows the different plots that can be made in MATLAB. We only need to select the variables we wish to plot. In the APPS tab we have the apps available in the toolboxes. We can also add the apps that we develop.

1.3   Simple Calculations in MATLAB

MATLAB can perform simple calculations as if it were a simple calculator. The prompt symbol >> indicates that MATLAB is ready. For example, if we wish to add two numbers such as 2 + 3, we simply write in the Command Window after the MATLAB prompt the desired operation as:

```
>>  2 + 3
```

and press ENTER, the output that MATLAB gives is

```
ans =
 5
```

We use the convention that the data we enter in MATLAB and the results from MATLAB are written in typewriter font.

We now show in Table 1.1 the basic operations and the precedence they have in MATLAB. The precedence means the priority order to perform the different operations. So, exponentiation has the highest priority, multiplication and division have the following higher priority, and finally, addition and subtraction have the lowest priority. Thus, for example, in the operation

```
>>  2*3 + 4*7∧4
```

The exponentiation 7∧4 is performed first, then the multiplication 2*3 and 4 times the result of 7∧4, and finally, the addition of the results of 2*3 and 4*7∧4.

Precedence can be user-modified by using parentheses. Operations enclosed by parentheses are performed first. For example,

```
>>  2∧3*4
```

TABLE 1.1: Basic MATLAB operations and precedence

Operation	Symbol	Example	Precedence
Addition	+	15 + 8 = 23	3
Subtraction	-	16 - 11 = 5	3
Multiplication	*	3*7.2 = 21.6	2
Division	/	18/3 = 6	2
Exponentiation	∧	3∧4 = 81	1

gives the result

```
ans =
   32
```

Since the first operation is 2∧3, which gives 8, then this result is multiplied by 4. Instead, we can use parentheses as in

```
>>  2∧(3*4)
```

and we obtain

```
ans =
   4096
```

because MATLAB performs first (3*4) which gives 12 and then 2∧12 which gives 4096.

1.3.1 Elementary Functions

MATLAB has a set of functions for everyday use. These functions are available in a set called **Elementary Functions**. The trigonometric functions fall within this set. For example, to calculate **sin(1)**

```
>>  sin(1)
```

```
ans =
   0.8415
```

For the case of the trigonometric functions, the argument is in radians. Some of the elementary functions are available in Table 1.2. For a listing of all of the elementary functions, we enter **help elfcn** in the **Command Window**.

TABLE 1.2:   Elementary functions

Function	MATLAB Notation
sin x	sin (x)
cos x	cos (x)
tan x	tan (x)
$\sqrt{x}$	sqrt (x)
$\log_{10}(x)$	log10 (x)
ln (x)	log (x)
$|x|$	abs (x)
e^x	exp (x)

The word `elfcn` is an acronym for **elementary functions**. As mentioned above, MATLAB has more than ten thousand functions. Some of the most elementary ones are written in C and most of the functions are written in the MATLAB language.

For the case of trigonometric functions, a simple example is to find the sine of the irrational number π. If we approximate π by 3.1416, then

```
>> sin (3.1416)
```

```
ans =
   -7.3464e-006
```

which is a good approximation to the exact result. MATLAB has a predefined value for the constant π and which is simply called **pi**. Then, we have

```
>> sin(pi)
```

```
ans =
   1.2246e-016
```

which is closer to the expected result. Other examples are

```
>> sqrt (2)
```

```
ans =
   1.4142
```

```
>> log10 (1000)
```

```
ans = 3.0000
```

Some of the predefined constants are shown in Table 1.3.

TABLE 1.3:   Predefined constants in MATLAB

MATLAB	
pi	3.14159265
i	Imaginary unit $= \sqrt{-1}$.
j	Same as i.
eps	Precision of the floating point operations, $2.22*10^{-16}$
Inf	Infinity
NaN	Not a number.

1.4   Variables

Variables are created in MATLAB as they are defined. That is, there is no need to predefine them together with their type as it has to be done in languages such as C, Fortran, and Visual Basic, to name a few. For example, the variable `alpha` is created the first time we use it as:

```
>> alpha = 32

alpha =
    32
```

Now, this variable appears in the `Workspace` window and will remain there until we delete it with the instruction `clear` which erases all variables stored in the `Workspace`. To see a list of variables in the `Workspace` we only need to type `who`. For example, if the MATLAB session just started, the only variable is `alpha`. Then, we obtain

```
>> who

Your variables are:
alpha
```

Variable names can have up to 63 characters. If a variable name is longer than 63 characters, the name will be truncated to 63 characters.

The number of digits used by MATLAB can be changed according to the format chosen for this purpose. The formats available are shown in Table 1.4. We show the formats with the irrational number $\sqrt{2}$.

Each time MATLAB performs an action, the result is displayed in the `Command Window`. Sometimes we are only interested in the final results and not in the intermediate ones. We can suppress intermediate results by typing a semicolon after the instruction at the end of the line. For example:

TABLE 1.4:   Formats for displaying numerical values

MATLAB format	Displayed value	Comments
format short	1.4142	5 digits
format	1.4142	Same as format short
format long	1.414213562373095	16 digits
format short e	1.4142e+00	5 digits plus exponent
format long e	1.414213562373095e+00	16 digits plus exponent
format hex	3ff6a09e667f3bcd	Hexadecimal
format bank	1.14	2 decimals (for currency)
format +	+	Positive or negative
format rat	1393/985	Approximates to a ratio

```
>>  23 + 32

ans =
   55

>>  23 + 32;
```

The first time we calculate 23 + 32 we get an answer, the second time we have a semicolon after the indicated addition and we **do not** get an answer at all; however, the result is stored in the variable **ans**.

MATLAB variable names are case sensitive, so care must be taken when writing long programs as we see in Chapter 6. Thus, variable A1 is different from variable a1.

For a set of numbers written between brackets as in

```
v = [ pi 2 -4 8 sin(2)]
```

The variable v is called a vector, in this case a row vector. In a column vector, the elements are separated by semicolons, thus it is written as

```
w = [2 ; 3; -7; sqrt(3)]
```

In the column vector w the elements are separated by a semicolon. This indicates that the elements after the semicolon are located in a different row. Thus, MATLAB gives as an answer to these two vectors:

```
>>  v = [ pi 2 -4 8 sin(2)]

v =

   3.1416 2.0000 -4.0000 8.0000 0.9093
```

```
>> w = [2 ; 3; -7; sqrt(3)]

w =
2.0000
3.0000
-7.0000
1.7321
```

In the case of row vector v, the elements can be separated by either blanks or by commas, while in a column vector elements are separated by semicolons. A vector has dimension n if it has n elements. Thus, vector v has dimension 5 whereas vector w has dimension 4.

If we have two row vectors with the same dimension, we define the dot product by

$$x*y'$$

but if we have two column vectors with the same dimension, we define the dot product by

$$x'*y$$

The result is a scalar. Thus, for vectors v, w given by

```
>> v = [-1 4 7 -9]

v =
   -1 4 7 -9

>> w = [9 8 -6 3]

w =
    9 8 -6 3

>> v*w'

ans =
   -46
```

1.4.1 Variable Types

Variables have a type in MATLAB. The available types are shown in Table 1.5 and we have also the size in bytes that each type uses in the memory.

TABLE 1.5:   Data types

Data type	Name	Size
double	double precision	8 bytes
single	single precision	4 bytes
int8	8-bit signed integer	1 byte
int16	16-bit signed integer	2 bytes
int32	32-bit signed integer	4 bytes
int64	64-bit signed integer	8 bytes
uint8	8-bit unsigned integer	1 byte
uint16	16-bit unsigned integer	2 bytes
uint32	32-bit unsigned integer	4 bytes
uint64	64-bit unsigned integer	8 bytes

1.5   Strings

A string of characters is a series of letters, and/or numbers, and/or any other symbols. A string is defined by enclosing the set of characters between single quotes. As examples of strings we have 'book', 'technical computing', '2332'. A variable in MATLAB can be a string. For example,

```
>> a = 'taylor'

a =
taylor

>> b = '1 a 2 b 3 c!"#'

  b =
        1 a 2 b 3 c!"#
```

If for any reason a string has to have a quote as a part of the string, we simply repeat the quote as in

```
>> a = 'Laura''s bear'

a =
   Laura's bear
```

In this example, the blank space is also a character in the string.

The elements in a string have a position. Thus a(k) is the kth element in the string. The number k is called the index. For the string a given above, a(1) = L and a(10) = e,

```
>> a(1)

ans =
   L

>> a(10)

ans =
   e
```

To specify a range of characters we use a(k:n) which indicates the characters from the *k*th to the *n*th. For example,

```
>> a(3:9)

ans =

   ura's b
```

We can also use a(3:2:9) which indicates the set of characters in a string from index 3 to index 9 but incrementing the index by 2, that is, we are looking for a(3), a(5), a(7) and a(9),

```
>> a(3:2:9)

ans =

   uasb
```

We can concatenate several strings by using

```
        New_string = [string 1, string 2, string 3]
```

For example, if we have three strings a1, a2, and a3, defined by (there is a blank space after each country name)

```
>> a1 = 'Canada '; a2 = 'Mexico '; a3 = 'USA ';
```

We can form

```
>> b = [a1, a2]
```

TABLE 1.6:   Functions for strings

MATLAB	Function
length	Number of characters in a string.
strcmp	Compares two strings.
str2num	Converts a string to a numerical value.
num2str	Converts a number to a string.
strrep	Replaces characters in a string with different characters.
upper	Changes lowercase characters to uppercase.
lower	Changes uppercase characters to lowercase.

```
b =

Canada Mexico

>> b2 = [a1, a3]

b2 =
Canada USA
```

We can also form new strings with elements from the strings already defined as

```
>> y1 = [a1(1:3), a2(1:3), a3(1:2)]

y1 =
CanMexUS
```

There exists a set of functions that can be performed on strings. Table 1.6 shows a list of some of the most used ones.

For the function length and the string a = 'This is Chapter 1', we have

```
>> length(a)

ans =
17
```

The instruction strcmp(a, b) compares two strings a and b. If they are equal the result is unity, if they are not equal the result is zero. For example, for the strings a1 and a2 given above,

```
>> strcmp(a1, a2)
```

```
ans =
   logical
   0

>> strcmp(a2, 'Mexico ')

ans =
   logical
   1
```

The instruction `str2num` converts a string to a number only if the string characters are numbers, as

```
>> d = str2num('2010')

d =
2010
```

The instruction `num2str` transforms a number into a string. For the number a = 810

```
>> num2str(a)


ans =
   '810'
```

which is a string. The instruction `strrep` (c1 , c2 , c3) replaces the string c2 by string c3 in the string c1. For example, in the string 'MATLAB is a high level programming language' we can replace the string 'level' by 'level technical' with

```
>> c1 = 'MATLAB is a high level programming language' ;
>> c2 = 'level';
>> c3 = 'level technical';
>> strrep( c1, c2, c3)


ans =
   'MATLAB is a high level technical programming language'
```

The instructions `upper` and `lower` convert lowercase characters to uppercase and vice versa, respectively. For the strings a = 'American Continent' and b = 'European Continent' we have:

```
h = upper(a)
k = upper(b)
lower(h)
lower(k)

h = 'AMERICAN CONTINENT'
k = 'EUROPEAN CONTINENT'
ans =
'american continent'
ans =
'european continent
```

1.6   Saving a Session and Its Variables

A variable and its value are kept only during the session and only until the user uses the instruction `clear` which deletes all the variables in the session. However, in several cases it is desired to have them available in another session. In other cases, we wish to save not only the variables, but also the instructions executed during the session. That is, we might need to keep the complete session. This can be done using the instruction `diary` followed by the file name. The instructions and variables will be stored in that file until the instruction `diary off` is executed, as in the following,

```
diary file_name
   ⋮
diary off
```

where the dots indicate the instructions and variables. Let us consider the following session

```
>> a = 1; b = 2;
>> diary session.txt
>> b = 3;
>> a = b∧2

a =
    9.0000

>> x = -b/a

x =
   -0.3333
```

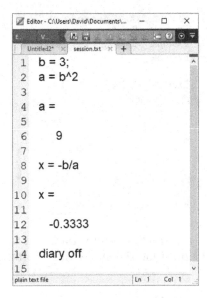

FIGURE 1.2: Instructions and variables saved with `diary`.

```
>>  diary off

>>  d = 3*x

d =
   -1.0000
```

Now we take a look at the `Current Folder` window and we observe that there is a new file called `session.txt`. If we open it with `File → Open` we see that all the variables and instructions together with the responses executed after `diary session.txt` and before `diary off` are stored in the file. This is shown in Figure 1.2. If a file name is not specified, the default file name is `diary.m`.

Another way to save information is by using the icon `Save Workspace` which is available from the tools bar. In this case we **only** save variable information. We cannot save instructions with this option. In this case the file name has the extension `.mat`. For this session we save the data in the file `sessionOne.mat`.

Alternatively, we can also execute

```
>> save('sessionOne')
```

To load the variables in a later session we simply execute

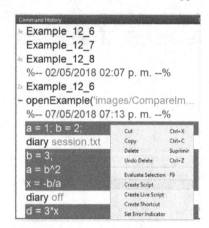

FIGURE 1.3: Selecting instructions and variables in the `Command History` window using the mouse right button.

>> load ('sessionOne')

We can also use icon **Open** and select the file variables.

From the `Command History` window we can also save information. By selecting with the left button mouse the instructions and variables and then with a right click we select the option **Create Script** as shown in Figure 1.3. The file name is saved with the extension .m. The final file is shown in Figure 1.4.

1.7 Input/Output Instructions

Up to now we have used MATLAB as a calculator, where the result is written right after we hit the ENTER key. Sometimes this is not very convenient, especially if we are not interested in intermediate results but rather at the final one. Fortunately, we can omit intermediate outputs by typing a semicolon after the instruction, as we saw in Section 1.4. Now we see how to read and write data from a to a file, respectively.

1.7.1 Formatted Output

Besides the direct output obtained after entering an instruction and hitting the ENTER key, we can use the instruction `fprintf`. We illustrate its use with an example. To write the string 'This is chapter 1.' we use:

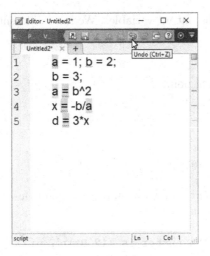

FIGURE 1.4: Instructions and variables saved in the file `variables.m`.

```
>> fprintf('This is chapter 1.\n' );

This is chapter 1.
```

We note several things. First we added \n which indicates that there is a line feed after writing the desired string. If we now use again the `fprintf` instruction, the string included there will be written in the following line. Second, the semicolon does not preclude writing the desired output. And third, there is not a line with **ans** as before. If we continue writing data with `fprintf` but we do not use \n, the output is going to be written in the same line. This is so because the instruction `fprintf` does not make a line feed by itself, but we have to indicate it. For example,

```
>> fprintf('This ');fprintf('is ');fprintf('Chapter 1.');

This is Chapter 1.

>> fprintf('This \n');fprintf('is \n');fprintf('Chapter 1.\n');

This
is
Chapter 1.
```

The first time we executed the `fprintf` instruction, the strings were written in the same row because we omitted the line feed part. In the second `fprintf` instruction we placed an \n before ending each string. Thus, after each `fprintf` instruction there is a line feed.

We can format the output variables. We have different formats to do this. The formats are:

f - floating point
g - fixed or floating point
i - integer
c - character
s - string
d - double precision
e - exponential notation

We can also use `fprintf` with numerical quantities. For example, for the number a = pi, we can use

```
>> fprintf ( 'The number is %12.8f\n', pi )
```

```
The number is 3.14159265
```

which means that the number `pi` is written as a fixed point number with 8 decimal places.

1.7.2   Data Input

The easiest way to input data to MATLAB is through the keyboard, in the same way we have done it so far. Another way is to use the instruction `input` which has the format

```
x = input ('Enter the value of x')
```

This instruction displays the string between quotes and waits for the value of x. So we can have the following:

```
>> x = input ( 'Enter the value of x ' ) ;
```

```
Enter the value of x  56
```

```
>> fprintf ( '\n You entered the number %g.\n ', x)
```

```
You entered the number 56.
```

The instruction `input` does not generate a line feed. As mentioned before, the line feed can only be generated with \n, which can go anywhere within the string.

If we enter an alphanumeric value instead of a numerical one, we get an error message as in:

```
>>  x = input ( 'Enter the value of x ' )

Enter the value of x

fifty-six

??? Error: Unexpected MATLAB expression.

Enter the value of x
```

Here we enter the text `fifty six` and get back an error message. MATLAB requests again the value of `x`. It keeps doing the same thing until we enter a numerical value.

To enter a string we add 's', as in:

```
>> x = input('Enter the value of x ', 's')

   Enter the value of x

fifty-six

x =
fifty-six
```

Since every input can be seen as a string, we never get an error message in this case.

1.8   Help

Help in MATLAB can be retrieved from the `Command Window` by typing `help` to obtain a list of topics.

```
>>  help


My Documents\MATLAB - (No table of contents file)
matlab\general - General purpose commands.
matlab\ops - Operators and special characters.
matlab\lang - Programming language constructs.
matlab\elmat - Elementary matrices and matrix manipulation.
```

```
matlab\elfun - Elementary math functions.
matlab\specfun - Specialized math functions.
matlab\matfun - Matrix functions - numerical linear algebra.
matlab\datafun - Data analysis and Fourier transforms.
matlab\polyfun - Interpolation and polynomials.
matlab\funfun - Function functions and ODE solvers.
matlab\sparfun - Sparse matrices.
matlab\scribe - Annotation and Plot Editing.
    :
```

For more help on directory/topic, type ''help topic''.
(Only the first few lines are displayed here).

To obtain help for a particular subject, type `help` followed by the function name. For example, to obtain information about the elementary functions we type `help elfun` to obtain a list of all the functions in the set `elfun`.

```
>> help elfun
```

To obtain information about a specific function, we type `help` followed by the function name. If we wish to get information about the sine function we use

```
>> help sin
```

```
SIN Sine of argument in radians. SIN(X) is
the sine of the elements of X.
See also asin, sind.
```

1.8.1 Help Page

We can also get help from the `Help` window. This window is opened by clicking on the `Help` icon in the `Tools` bar. We can also access this page by typing help in the `Command Window`. After doing this we obtain the window shown in Figure 1.5. In this window we can see a list which contains the topics. There is a `Search Documentation` window that has a field for data. There we write the instruction name or part of it to look for every instruction which has that name. The information is displayed in the right-hand window. Figure 1.6 shows an example for the function `diff` which calculates the symbolic derivative of a function.

FIGURE 1.5: Help window.

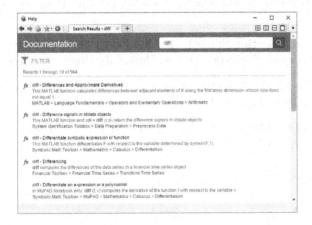

FIGURE 1.6: Search results for the function `diff`.

1.9   Concluding Remarks

We have started using MATLAB in this chapter. In the following chapters we will cover how it can be used to perform an unlimited number of tasks, going from the very obvious ones such as an addition to very complicated ones such as optimization of a product or process. MATLAB is used in education, research, and testing to name a few applications. We have only described the environment where all those calculations can take place.

In this chapter, we only have covered the way the several types of variables are defined and stored, both numeric ones and strings. Strings are a type of non-numeric variable which finds use in the creation of text for input and output information and for names of variables. We have also described the way to format data for input and output instructions.

Finally, we covered how a session, including variables and instructions, can be saved for later use.

1.10   Bibliography

1. A. Gilat, MATLAB: An Introduction with Applications, J. Wiley and Sons, New York, 2016.

2. D. C. Hanselman, Mastering MATLAB, Prentice Hall, Inc., Piscataway, NJ, 2008.

3. R. Pratap, Getting Started with MATLAB: A Quick Introduction for Scientists and Engineers, Oxford University Press, New York, 2016.

4. B. Hahn and D. Valentine, Essential MATLAB for Engineers and Scientists, Third Edition, Butterworth-Heinemann, Burlington, MA, 2016.

5. H. Moore, MATLAB for Engineers, 5th Edition, Prentice Hall, Inc., Piscataway, NJ, 2017.

Chapter 2

Plotting with MATLAB

2.1 Introduction

A very useful and powerful MATLAB characteristic is that very high quality and very complex plots can be done very easily. Thus, two-dimensional and three-dimensional plots are possible to do. This aspect makes data visualization possible in mathematical procedures. In this chapter we cover the instructions to produce many of the plots available in MATLAB. Also, we see the way the properties for each plot can be changed.

The chapter is organized as follows: first we treat the several one-dimensional and two-dimensional plots as well as the way their characteristics can be changed. Then, we move to three-dimensional plots. The next topic we introduce is the concept of handles, which is not exclusive to but mostly used with plots. Finally, we give a brief but detailed coverage of object hierarchy in MATLAB.

2.2 Two-dimensional Plotting

The basic instruction to plot a function is plot(X, Y). Here X and Y are vectors with the information about the independent and the dependent variables. We generate first the vector X and then we evaluate the function we wish to plot at the points of vector X to find vector Y. With the instruction plot, MATLAB generates a new window called Figure 1 where the plot is displayed. The vector X can be generated with the instruction linspace as in

$$x = \text{linspace(x\_1, x\_2, n)}$$

This produces a set of points equally spaced. The first point is x_1, the last point is x_2, and the distance between points is $(x\_2 - x\_1)/(n-1)$. As an example,

```
>> x = linspace(0, 5, 6)
```

produces the set of points given by

```
x =
      0  1  2  3  4  5
```

We see that the first point is 0, the last point is 5 and there are 6 points. The distance between consecutive points is $(5 - 0)/(6 - 1) = 1$. Then, the points are stored in a row vector of dimension 6. If we now wish to plot the function cos(x) from 0 to 4π with 200 points between these limits, we use

```
>>  x = linspace(0, 4*pi, 200);
>>  y = cos(x);
>>  plot(x, y)
```

We obtain the plot shown in Figure 2.1. We can also use the plot instruction with:

```
>>  x = linspace(0, 4*pi, 100);
>>  plot( x, cos(x));
```

and we obtain the same result as Figure 2.1. We also see that this window is labeled Figure 1 in the upper left corner. Also note that there is a tool bar above the plot. The tool bar has, besides the regular icons as New, Open,

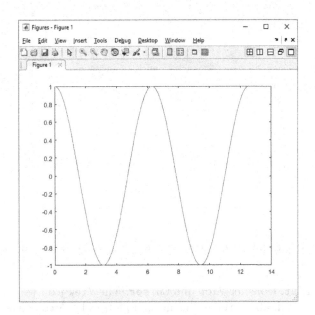

FIGURE 2.1: Plot of cos(x).

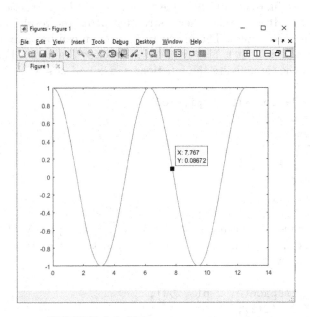

FIGURE 2.2: Using a cursor in a plot.

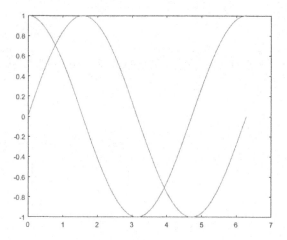

FIGURE 2.3: Two curves in the same plot using `hold on`.

`Save`, and `Print`, icons for `View` and `Cursor`. The `View` icon will be discussed later when we talk about three-dimensional plots. The `Cursor` icon is used to see the coordinates of a particular point in the plot. We just select the icon and place the cursor on top of the curve and the coordinates of that point will be displayed next to the cursor. An example is shown in Figure 2.2. If we wish to plot two functions on the same figure, we use the instruction `hold on`, as shown below. If we use another instruction `plot`, the resulting curve will appear in the same figure. The instruction `hold on` can be canceled with an instruction `hold off`. It is also canceled if we close the figure. For example, let us plot in the same figure the functions `sin x` and `cos x`:

```
>>  x = linspace(0, 2*pi, 100);
>>  y = cos(x);
>>  plot(x, y);
>>  hold on
>>  z = sin(x);
>>  plot(x, z)
>>  hold off
```

The resulting plot is shown in Figure 2.3. We can also plot two functions by defining pairs `x`, `y`. For example, to plot the functions `sin 2x` and `cos 3x` in the same figure we use:

```
>> x = linspace(0, 2*pi, 100);
>> y1 = sin(2*x);
>> y2 = cos(3*x);
>> plot(x, y1, x, y2)
```

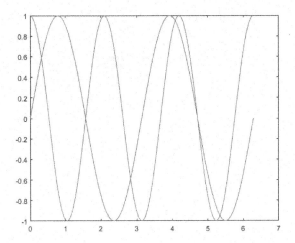

FIGURE 2.4: Two traces in the same figure using pairs x, y.

The results are shown in Figure 2.4. Note that each trace has a different color. We can add text information to the plot using the instructions xlabel, ylabel, title and legend. They add text to the x-axis, y-axis and to the plot, respectively. For our plot we can add:

```
>> xlabel('x-axis')
>> ylabel('y-axis')
>> title('Functions sine and cosine');
>> legend('sin(x)', 'cos(x)')
```

The figure now looks as shown in Figure 2.5. Other plots may have a semilogarithmic axis, either for the x-axis or the y-axis. For a semilog x-axis we use a semilogx and for a semilogarithmic y-axis we use a semilogy instruction as in

```
>> x = linspace(0, 2*pi, 100);
>> y = sin(x);
>> semilogx(x, y)
```

The plot is shown in Figure 2.6. Note that the x-axis is in a logarithmic scale while the y-axis is linear. For a semilog y-axis we use instead semilogy as in

```
>> x = linspace(0, 2*pi,100);
>> y = exp(x.^2);
>> semilogy(x, y)
```

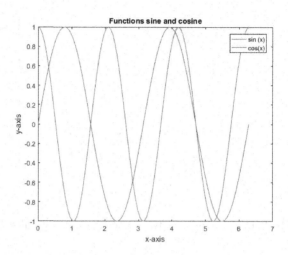

FIGURE 2.5: Text information added to plot of `sin x` and `cos x` functions.

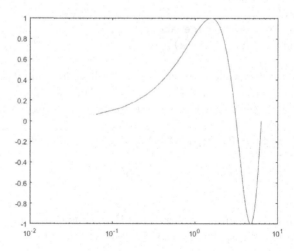

FIGURE 2.6: Semilog plot for `sin(x)`. The x-axis is in a log scale.

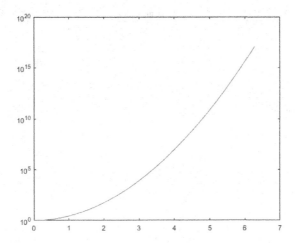

FIGURE 2.7: Semilog plot for `exp(x^2)`. The y-axis is in a log scale.

The plot is shown in Figure 2.7. Another way to have several functions in the same plot is to form a vector of functions and plot this vector. In this case the vector of x points must be a column vector. For example, to plot the functions `sin(x)`, `cos(x)`, `sin(x)*cos(x)`, `x*sin(x)`, we proceed in the following way:

```
>> x = linspace(0, 2*pi, 100)';   % Transpose to make it a
                                   % column vector.
>> y = [sin(x), cos(x), sin(x).*cos(x), abs(x).*sin(x)];
>> plot(x, y)
>> xlabel('x-radians')
>> ylabel('Four functions')
>> title('Multiple plot')
>> legend('sin(x)', 'cos(x)', 'sin(x)*cos(x)', 'abs(x)*sin(x)')
>> grid on
```

The plot is shown in Figure 2.8. We have added an instruction `grid on` to generate a grid on the plot. The grid can be disabled with the instruction `grid off`.

The number of points in the vector x determines the plot smoothness. If the number of points in the trace is small, the plot could be a rough approximation to the true trace. For example, if the number of points is very small, as in

```
>> x = linspace(0, 2*pi, 10)';
```

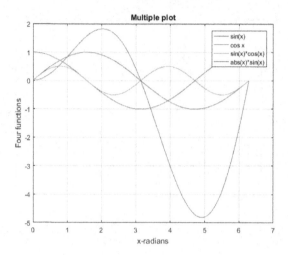

FIGURE 2.8: Multiple plot.

which produces a vector with only 10 points. The previous plot with this vector x is shown in Figure 2.9. Thus, it is better to have as many points as possible in the x vector.

2.2.1   Plotting from the Workspace

MATLAB allows to produce a plot directly from the Workspace. To see how this is done, let us consider a new MATLAB session. Now we generate the vectors x and y as follows:

```
>> x = 0:0.01:10;
>> y = exp(x).*sin(2*pi.*x);
```

We now take a look at the Workspace window and there we see icons for the variables just created as shown in Figure 2.10. Also we see that the Workspace window has a tool bar with some icons disabled. Now we select the variable y which corresponds to data for the function `exp(x)*sin(x)` and as soon as we right click on it a pull-down menu displays (see Figure 2.11). If the plot we wish to produce does not appear in the pull down menu, then we select `Plot Catalog...` and a dialog window in Figure 2.12 is displayed and there we look for the `plot(y)` option. When we click on it the plot is produced as shown in Figure 2.13. The `Plot Catalog` shows the different plots that can be produced. We will cover those other plots in the following sections.

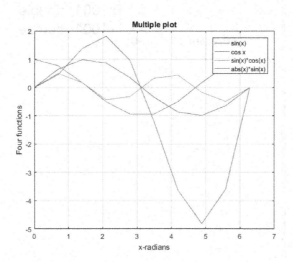

FIGURE 2.9: Plot with a few points for the x-axis.

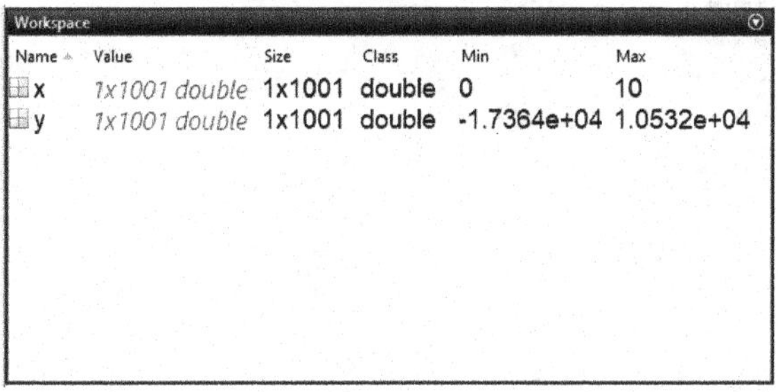

FIGURE 2.10: Workspace window with variables x, y created.

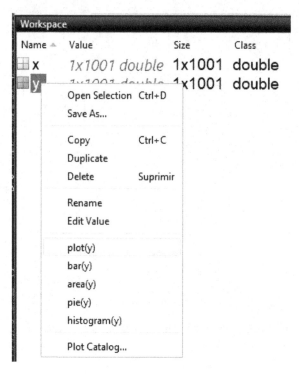

FIGURE 2.11: Selection of the plot type.

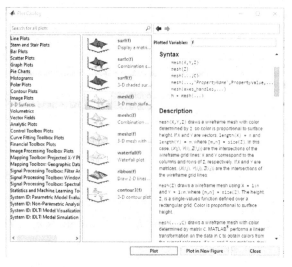

FIGURE 2.12: Plot catalog window.

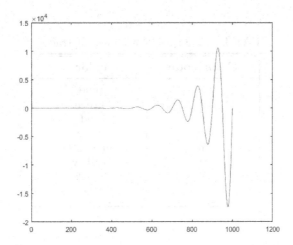

FIGURE 2.13: Plot for the variable y.

FIGURE 2.14: Icon for editing the figure properties.

2.3 Plot Options

There are several options we can use to give more information or make it more informative to users. The general form of the instruction plot is

```
plot(X1, Y1, S1, X2, Y2, S2,...)
```

where Xi, Yi are the vectors we wish to plot and Si has the information about the traces, such as the color, width, form, etc. The terms Si are strings containing the information about the options required by the user. This information is available in Tables 2.1–2.3. In these tables we find the strings where we can specify color, markers, marker size, trace width, and trace styles. These options can also be changed with the property editor. To use this editor, we click on the Edit Icon, shown in Figure 2.14 and this action allows us to select the element of the figure we wish to edit. To do this we just right click and select the parameter we wish to change. In this menu we can change line width, marker, color, marker size, and line style.

TABLE 2.1:   Color codes for traces

Color code	color
b	blue
g	green
r	red
c	cyan
m	magenta
y	yellow
b	black
w	white

TABLE 2.2:   Trace markers

Marker code	marker
.	point
o	circle
x	cross
+	plus sign
*	asterisk
s	square
d	diamond
V	downward pointing triangle
∧	upward pointing triangle
<	left pointing triangle
>	right pointing triangle
p	five point star
h	six point star

TABLE 2.3:   Trace style

Style code	trace style
-	Solid line
:	points
-.	Dash and point
- -	Dashed line

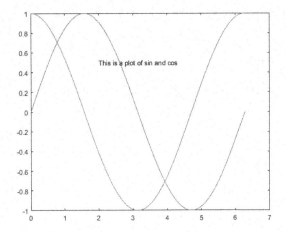

FIGURE 2.15: Text on a plot.

When we need to draw several traces, the colors are different for each one. The colors are selected by default starting with blue and continuing downward in Table 2.1. If the background is white, this color is not used for the traces. The background color can be changed with the instruction whitebg, for example:

```
>> whitebg('blue')
```

changes the background color from white to blue. We can go back to the previous color with the same instruction.

To place text on a plot we use the instruction text with the following format:

```
>> text( x, y, 'string')
```

x, y are the coordinates where we start the text on the plot and string with the text. For example,

```
>> x = linspace( 0, 2*pi, 100);
>> x = linspace( 0, 2*pi, 100);
>> y1 = sin(x);
>> y2 = cos(x);
>> plot( x, y1, x, y2)
>> text( 2, 0.5, 'This is a plot of sin and cos')
```

and we get the plot shown in Figure 2.15.

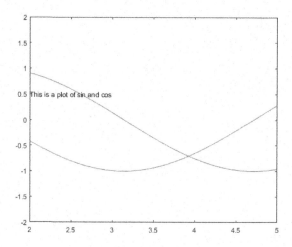

FIGURE 2.16: Change of axes' limits.

By default, MATLAB uses the information from vectors x, y to assign upper and lower bounds in the axes. However, the user can adjust them with the instruction axis that has the following format:

axis([x_initial x_final y_initial y_final])

For example, for our last plot we can use

>> axis([2 5 -2 2])

This instruction specifies that variable x goes from +2 to +5 and variable y goes from -2 to +2. The resulting plot is shown in Figure 2.16.

Another way to make changes in the axes limits in a plot is through the use of the figure's menu. From the figure's **Edit menu** select **Edit → Axes Properties...** which opens the **Axis** menu below the figure as shown in Figure 2.17. In this menu we see two small zones. In the left zone we can change the color properties to the different components of the plot such as the **fonts**, **traces**, **grid**, **background**, and we are also allowed to enter the text for the figure's title. In the right zone we can change the x, y, and z axes limits, change scale from **linear** to **log** and vice versa, add a text to the axes, as well as to change the font for each axis information.

2.4   Other Two-dimensional Plots

In this section we present the different two-dimensional plots available in MATLAB.

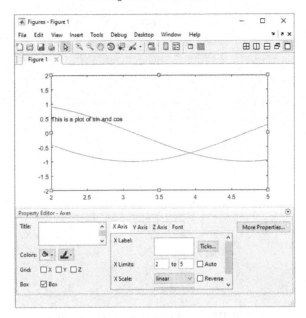

FIGURE 2.17: Plot with axis' limits menu displayed.

2.4.1 Polar Plots

We can make a polar plot if the function is written in polar coordinates. The instruction is

```
polar(θ, r, s)
```

where θ, r are the polar coordinates and s is a string with the same options as in the instruction plot. For example, if we wish to plot the function, given in polar coordinates by the equation

$$r = \theta \wedge 0.5$$

We use

```
>> theta = linspace(0, 8*pi, 200);
>> r = theta.∧0.5;
>> polar(theta, r)
```

This set of instructions produce the plot shown in Figure 2.18. We see that the function we plotted is a spiral. The dot before the sign $\wedge$ indicates that each element in the vector theta is elevated to the power 0.5.

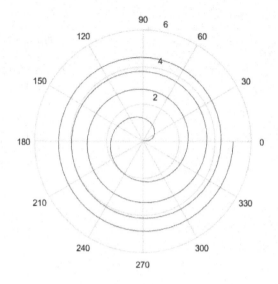

FIGURE 2.18: Polar plot of a spiral.

2.4.2   Bar Plot

A bar plot can be obtained with the instruction

$$bar(x, y, 's')$$

x, y are the vectors with the information and S is a string for the options. Note that it is the same format used by the instruction plot. As an example, to plot 21 points from -10 to 10 we use

```
>> x = linspace(-10, 10, 21);
>> y = sin(x);
>> bar(x, y);
>> title('Bar plot')
```

The bar plot is shown in Figure 2.19.

2.4.3   Stairs Plot

The same set of points can be plotted using a **stairs** instruction. The format is

$$stairs(x, y, 's')$$

In the previous plot, by changing **bar** for **stairs** and in the title instruction **bar** for **stairs** we get the **stairs** plot shown in Figure 2.20.

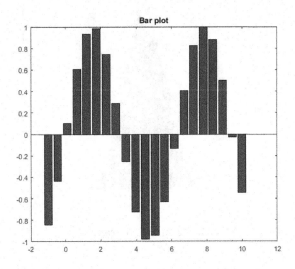

FIGURE 2.19: Bar plot.

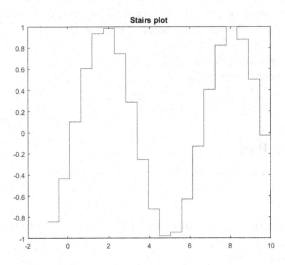

FIGURE 2.20: Stairs plot.

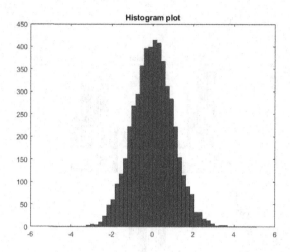

FIGURE 2.21: Histogram plot.

2.4.4 Histogram Plot

An histogram plot is similar to a bar plot, but we only get 10 bars with an histogram instruction. The instruction is

$$\text{hist(y, n)}$$

where y is the function to plot and n is the number of bars in the plot (the maximum number of bars is 10). For example, to get an histogram for a random variable we do the following to get Figure 2.21:

```
>> x = linspace(-5, 5, 50);
>> y = randn(5000, 1);
>> hist(y, x);
>> title('Histogram plot')
```

2.4.5 Stem Plot

A stem plot is a plot of points. The instruction is

$$\text{stem(x, y, 's')}$$

For example, for the set of points given by exp(-t)*sin(y) we get

```
>> x = 0:0.5:4*pi;
>> y = exp(-0.2*x).*sin(2*x);
>> stem(y,':');
```

This set of instructions produce the plot of Figure 2.22.

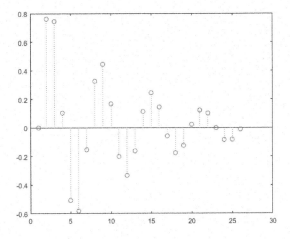

FIGURE 2.22: Stem plot.

2.4.6 Compass Plot

The instruction `compass` plots magnitude and phase of a set of vectors. The format is

$$\texttt{compass(z) = compass(x + jy) = compass(x, y)}$$

where `z` is a vector of complex numbers. For example, for vector `z` given by

$$\texttt{z = [3 + 2j, -4 + 7j, 6-9j, -4-5j]}$$

we get Figure 2.23 with

```
>> z = [3 + 2j, -4 + 7j, 6 - 9j, -4 - 5j]
>> compass(z)
```

2.4.7 Pie Plot

A pie plot is used to display percentages for each element in a vector or matrix. `pie` and `pie3` generate two-dimensional and three-dimensional plots, respectively. For the vector `A = [3 4 5 9 3]`, the `pie` plot is obtained with

```
>> A = [3 4 5 9 3];
>> pie(A)
```

and the result is shown in Figure 2.24.

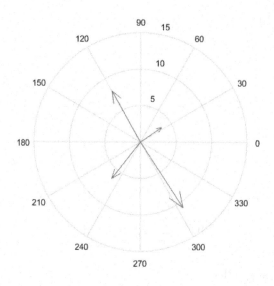

FIGURE 2.23: Compass plot.

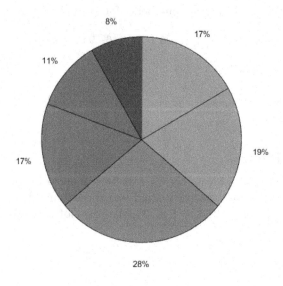

FIGURE 2.24: Pie plot.

2.5 Subplots

Sometimes we need to plot different functions in separate plots, but we need to visualize them at the same time. Fortunately, MATLAB allows us to do this by using subplots. Each plot in the figure is called a subplot. Each subplot is generated separately using the instruction `subplot` whose format is:

$$\text{subplot(m, n, p)}$$

This instruction divides the figure in m×n subplots arranged in a matrix form with m rows and n columns. The variable p activates each subplot. For example, for four subplots we use a 2×2 array and use `subplot(2, 2, p)` and p indicates the position in the array. In this case p can take values from 1 to 4. Thus, `subplot(2, 2, 1)` produces the plot in the 1, 1 position in the figure, `subplot(2, 2, 3)` produces the plot in the 2, 1 position in the figure, whereas `subplot(2, 2, 2)` corresponds to the 1, 2 position. For example,

```
>> subplot(2, 2, 1)
>> x = linspace(0, 2*pi, 100);
>> y = sin(x);
>> plot(x, y);
>> title('sine function')
>> subplot(2, 2, 2)
>> y = cos(x);
>> plot(x, y);
>> title('cosine function')
>> subplot(2, 2, 4)
>> r = 5*log10(x);
>> polar(x, r)
>> title('Spiral')
```

The subplots activated are located two in the first row and one in the second row, as shown in Figure 2.25. Since we did not generate a `subplot(2, 2, 3)`, there is an empty space in that position.

2.6 Three-dimensional Plots

Three-dimensional plots, besides being more attractive to viewers, give more information to them. There are several types of three-dimensional plots, some of them we cover in this section. For a three-dimensional plot there are three variables: x, y, z. We plot z which is a function of x, y, that is, z = f(x, y).

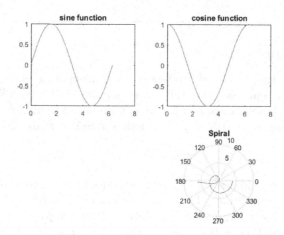

FIGURE 2.25: Multiple plots using the instruction `subplot`.

The three dimensions are then x, y, z. Depending upon which type of plot we need we use the appropriate instruction.

2.6.1   The Instruction `plot3`

The instruction to plot in three dimensions is `plot3`. It has the format

$$\text{plot3(x, y, z, s)}$$

Here, x, y, z are the coordinates and s is a string with the options for the plot. For example, the helix function has the coordinates:

```
>> x = sin t
>> y = cos t
>> z = t
```

Then, a three-dimensional plot can be obtained with

```
>> t = linspace(0, 10*pi, 500);
>> plot3(sin(t), cos(t), t);
>> title('Helix')
>> text(0, 0, 0, 'Origin')
>> grid on
```

The helix plot is shown in Figure 2.26.

Now let us suppose that we wish to plot three curves. These curves are defined by equations: sin x, cos x, and sin 2x, in three different planes.

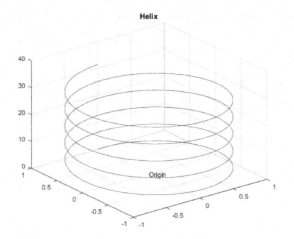

FIGURE 2.26: Helix plot using `plot3`.

This can be done if each function starts its values in a different value for the y variable, that is, in a different plane. The code to do this is:

```
>> x = linspace(0, 3*pi, 100);
>> Z1 = sin(x);
>> Z2 = cos(x);
>> Z3 = sin(x).*sin(x);
>> Y1 = zeros(size(x));
>> Y2 = ones(size(x));
>> Y3 = Y2/2;
>> plot3(x, Y1, Z1, x, Y2, Z2, x, Y3, Z3);
>> grid on
>> title('sin x, cos x, sin x^2')
```

After running this code we get Figure 2.27. There we see that each curve is plotted in a different plane.

2.6.2 Mesh Plot

A mesh plot is a three-dimensional surface with lines drawn forming a mesh over the surface. The functions that can be plotted are of the form

$$z = f(x, y)$$

and the instruction is

$$mesh(x, y, z)$$

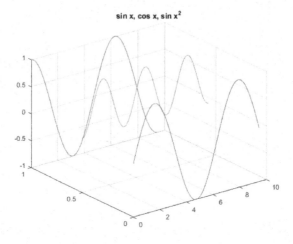

FIGURE 2.27: Three two-dimensional curves in a three-dimensional plot.

Before executing the mesh instruction, we need to create a matrix for x, y values. This can be done with the instruction meshgrid which has the format:

$$[x, y] = \text{meshgrid}(x\_i, y\_i: \text{inc}: x\_f, y\_f)$$

where (x_i, y_i) and (x_f, y_f) are the initial and final values for x, y, and the variable inc is the increment for x, y. For example,

```
>> [X, Y] = meshgrid(-10 : 0.5 : 10);
>> R = sqrt(X.∧2 + Y.∧2) + eps;
>> Z = sin(R)./R;
>> mesh(X, Y, Z)
```

produces the meshplot of Figure 2.28. As we can see in the screen monitor, the surface is plotted in color. This meshplot can be made transparent by using the instruction **hidden off**. After executing this instruction, the new plot is shown in Figure 2.29.

The instruction mesh(x, y, z) can accept plot options, as was the case of the plot instruction. For example, for a figure in black, we get Figure 2.30 if we execute the following:

```
>> mesh(X, Y, Z, 'Edgecolor', 'black')
```

Two instructions similar to mesh are meshc and meshz. meshc adds a contour map over the x, y plane and meshz adds a zero plane. If we change mesh for meshc and then for meshz we get the plots shown in Figures 2.31 and 2.32.

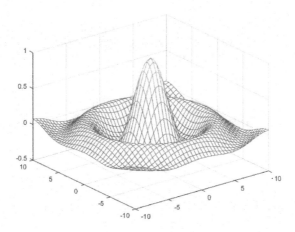

FIGURE 2.28: Mesh plot.

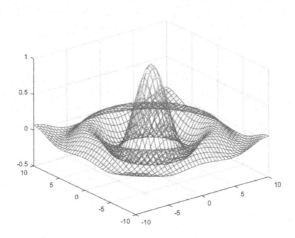

FIGURE 2.29: Transparent mesh plot using **hidden off**.

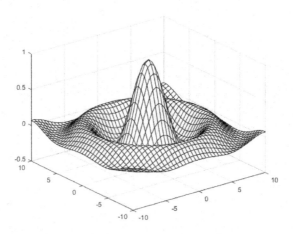

FIGURE 2.30: Black mesh plot using options.

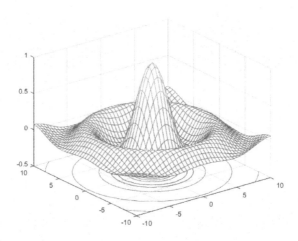

FIGURE 2.31: Mesh plot with a contour using `meshc`.

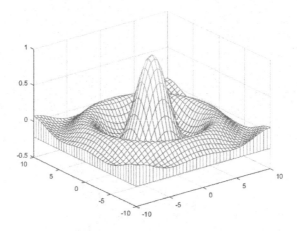

FIGURE 2.32: Mesh plot with a zero plane using `meshz`.

2.6.3   Surf Plot

The three-dimensional plot obtained with `surf` is similar to the mesh plot, except that the rectangular portions on the surface are colored. The colors are determined by the values of the vector `z` and by the color map. Let us consider as an example, the surface `sphere`, available from MATLAB. If we desire to plot this surface we use the following instructions:

```
>> [x, y, z] = sphere(12);
>> surf(x, y, z)
>> title('Sphere plot')
>> grid
>> xlabel('x-axis'), ylabel('y-axis'), zlabel('z axis')
```

To obtain the plot in Figure 2.33. In the sphere figure we can delete the black lines with the instruction

```
>> shading flat
```

to get Figure 2.34. If instead of **shading flat** we use the instruction:

```
>> shading interp
```

we obtain a plot with the colors smoothed, as shown in Figure 2.35. The instruction **surfl** produces a plot with lighting. It has the format

```
>> surfl(x, y, z)
```

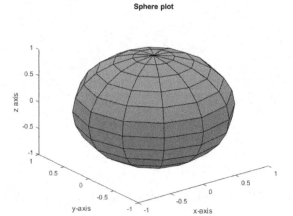

FIGURE 2.33: Plot of sphere with `surf`.

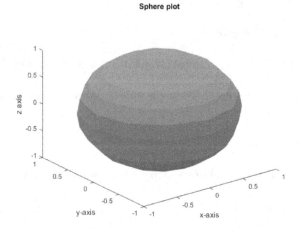

FIGURE 2.34: Plot of sphere without lines.

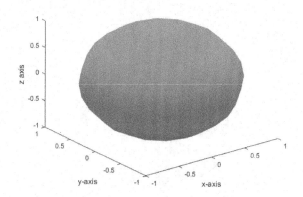

FIGURE 2.35: Plot of sphere with smoothed colors.

Using it for the sphere we produce Figure 2.36.

2.6.4 Contour Plot

The instruction `contour` gives a two-dimensional plot from a three-dimensional plot. The format is

```
>> [x, y, z] = peaks(30);
>> contour(x, y, z, 16) % contour with 16 colors
>> xlabel('x-axis'), ylabel('y-axis')
>> title('Peaks contour')
```

This set of instructions produces Figure 2.37. A variation of `contour` is `contour3` which produces a three-dimensional contour. If in the previous example we only change `contour` by `contour3` we obtain Figure 2.38.

A similar contour but using pseudocolor is obtained with the instruction `pcolor` as:

```
>> pcolor(x, y, z)
```

and the results are shown in Figure 2.39.

The instruction `waterfall`, also provides a contour plot but resembling a waterfall. For example, for the `peaks` function:

```
>> [x, y, z] = peaks(30);
>> waterfall(x, y, z )
>> xlabel('x'), ylabel('y'), zlabel('z')
```

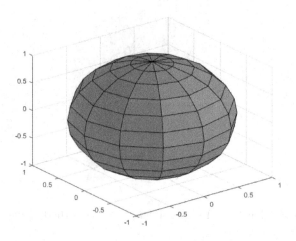

FIGURE 2.36: Surface plot with lighting.

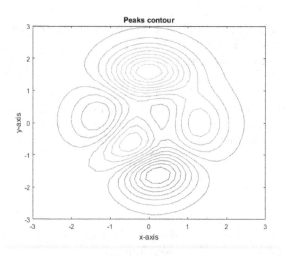

FIGURE 2.37: Contour plot.

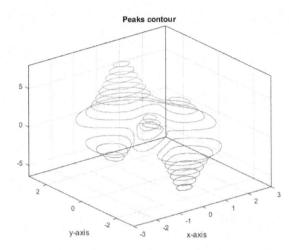

FIGURE 2.38: Three-dimensional contour plot.

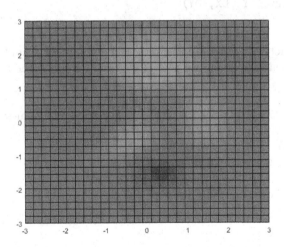

FIGURE 2.39: Three-dimensional contour plot with `pcolor`.

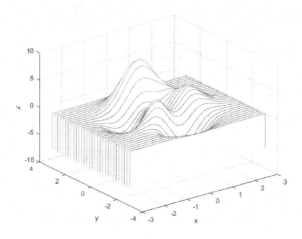

FIGURE 2.40: Waterfall plot.

These instructions produce the plot shown in Figure 2.40.

The instruction `quiver` gives directional lines to a `contour` plot. We have to define a differential vector `DX, DY` using a gradient instruction. For example,

```
>> [X, Y, Z,] = peaks(30);
>> [DX, DY] = gradient(Z, 0.5, 0.5);
>> quiver(X, Y, DX, DY)
```

which produces the plot in Figure 2.41.

The instruction `clabel` add values to the height of a `contour` plot. We have first to generate a contour plot. This can be done in the following way:

```
>> [x, y, z] = peaks(30);
>> cs = contour(x, y, z, 10); % For numerical values.
>> clabel(cs)
>> xlabel('x'), ylabel('y')
>> title('Contour plot for peaks with values')
```

The result is shown in Figure 2.42.

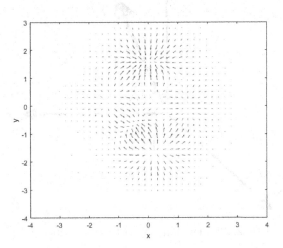

FIGURE 2.41: A quiver plot.

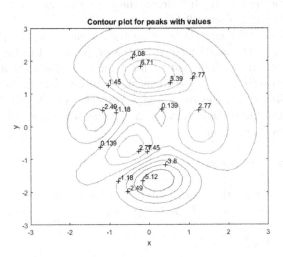

FIGURE 2.42: Contour plot with values.

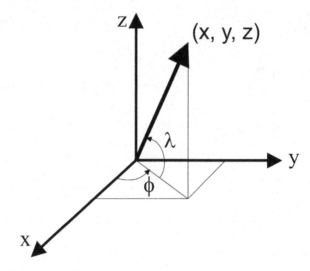

FIGURE 2.43: Viewpoint for a three-dimensional plot.

2.7   Observation Point

If we observe any of the three-dimensional plots from the previous sections, we see that the user sees them from a predetermined viewpoint. That viewpoint is, in spherical coordinates, the distance r, the azimuth angle ϕ and the elevation angle λ as is shown in Figure 2.43.

The elevation is the angle λ between the radius r and the plane xy. The azimuth angle ϕ is the angle formed by the projection of radius r on the xy plane and the x axis. Default values for azimuth and elevation angles are -37.5° y 30°, respectively. A two-dimensional plot has default values of 0° and 90° for the azimuth and elevation angles, respectively. The viewpoint can be changed with the instruction `view` by giving the new values for azimuth and elevation angles. The format is

```
view(elevation, azimuth)
```

For example, consider the mesh plot from Figure 2.28. We can change the viewpoint to an elevation angle of 30° and an azimuth angle of 60° with

```
>> view ( [ 30 60 ] )
```

to obtain Figure 2.44, which is the same mesh plot, but from a different viewpoint.

Alternatively, we can use the rotation button in the figure's toolbar. This

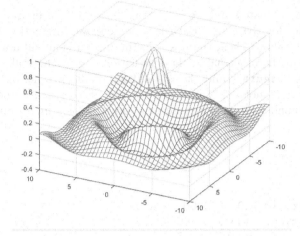

FIGURE 2.44: Mesh plot with a different viewpoint.

FIGURE 2.45: Rotate button.

button is shown in Figure 2.45. To activate the button, in the figure toolbar select **Rotate 3D** to display the rotae icon in the plot's window, just click once on the **Rotate** button and then position the cursor on the figure. Wait for a few seconds and then the azimuth and elevation angles will appear on the lower left corner of the figure. Then just move the mouse around the figure with the left-hand mouse button to move the plot.

2.8 Structure of Objects in MATLAB

If we take a closer look at the figures generated so far in the chapter, we see that all of them are Figure 1. This is so because each time MATLAB produces a plot, it creates a new figure window, if none is open, or it plots the new plot in the current figure, deleting whatever was plotted there (Unless there is an active **hold on**). To open several windows, each one with a different plot, we use the instruction **figure**. Its formats can be any one of the following:

```
figure
h = figure
figure(h)
```

A new figure is created and we call **h** the handle of the figure. The handle is an identifier for the figure. Each time we use the instruction **figure**, we open a new window. Then, the next plotting instruction will appear in that new figure. On the other hand, the instruction **figure(h)** makes active the window named **Figure h** and places it in front of all the other open windows. If **Figure h** does not exist, it is created and it is assigned the handle **h**. The handle number must be an integer in order to be a figure's number. For example,

```
>> f1 = figure
    Number: 1
    Name: "
    Color: [0.9400 0.9400 0.9400]
    Position: [403 246 560 420]
    Units: 'pixels'
    Show all properties
```

This creates a new window with handle value equal to 1 and named **Figure 1**. We can also see other properties for this figure. Now, if we repeat the instruction but using the semicolon to avoid the properties list as:

```
>> f2 = figure;
```

Another window is created, this time for **Figure 2** with **handle = 2**. Note that as we continue this process, the handles are increasing in value consecutively. Thus, 3 would be the next handle, and so on. We can also assign a handle to any figure. For example, if we wish that our next window is **Figure 5** we do the following:

```
>>  f3 = figure(5);
```

We now have three figure windows open for Figures 1, 2 and 5. Our next step is to draw some plots on them. We do this with the following set of instructions:

```
>>  x = linspace(0, pi, 25);
>>  y = sin(x);
>>  plot(x, y)
```

We see that the plot was drawn in **Figure 5** because this is the active window. This is shown in Figure 2.46. If we now write:

```
>> figure(f1)
```

The window corresponding to Figure 1 would be the active one and if we plot

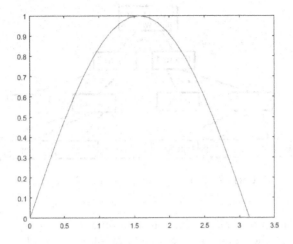

FIGURE 2.46: Plot in Figure 5.

a function, it appears in this window. For example, for the instructions:

```
>>  y2 = cos(2*x);
>>  plot(x, y2);
```

we see that this function is plotted in the window corresponding to Figure 1.

Other instructions useful for figure windows are: Instruction **shg** makes the active figure to be displayed on top of the other windows. The instruction **clf** deletes the active windows, while **close(h)** closes the window with handle h and **close** is used to close the active figure. The instruction textttclose all closes every figure window.

2.9 Hierarchy of MATLAB Objects

Windows, graphs, and every object in MATLAB have a hierarchy. They have a hierarchic structure due to the different types of objects in each of them. Such hierarchy has a tree-like form, as shown in Figure 2.47. According to the figure, the most general object is the main MATLAB window. This object is the root for all other objects. There is only a root object. The root object, the MATLAB window, can have several windows such as the **Workspace**, a figure window, a properties window, etc. Each of these windows can have objects such as axes, title, traces, menus, controls, etc. Finally, each of these objects can have below lower hierarchy objects such as lines, rectangles, patches, surfaces, and text. Some objects are called parents. The objects below a parent object

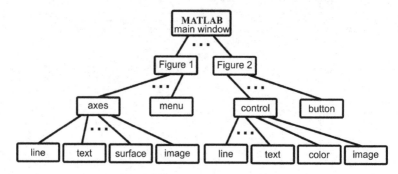

FIGURE 2.47: Hierarchy of MATLAB objects.

are called children of that object. Thus, an object can be a parent and have children, and at the same time be a children of a parent object with higher hierarchy. When an object is deleted from MATLAB, its children objects are automatically deleted too.

2.10 Concluding Remarks

It is very easy to produce high quality plots in MATLAB, even in the most complex cases. In this chapter we have covered several plots available in MATLAB. We have seen that they are easily produced. Plots in two and three dimensions have been obtained. In addition, we have seen how any plot is organized in MATLAB. A very important concept related to plots is the concept of handles. In the following chapters, we will use plots in a great deal of examples.

Chapter 3

Variables and Functions

3.1   Variables

MATLAB has a very powerful way to define variables. Most programming languages have to define variables before they are used for the first time. MATLAB does not need to do that. Each variable is defined as they are used. If in a procedure we need to define the variable Y0, it is defined the first time we use it. Thus, if we write

```
>> Y0 = 1
   Y0 =
      1
```

This variable has that value until we change its value or we delete it with the instruction `clear` as

```
clear Y0
```

A variable name can have up to 63 characters. If a variable name is longer than 63 characters, MATLAB keeps only the first 63 characters and eliminates the remaining ones.

Variable names have to start with a letter and can only have letters, numbers and underscores.

All the variables are real unless we define them as complex. The variable Y0 defined above is a real variable. To define a complex variable we use the imaginary number $i = j = \sqrt{-1}$. We can define a complex number as:

```
>> z = 2 + 3*j
   z =
      2.0000 + 3.0000i
```

In the case of complex numbers such as $c = a + b*i$, a and b are called the real and imaginary parts, respectively. They can be found with `real(c)` and `imag(c)`. For the complex number z we have

```
>> real(z)
   ans =
      2
>> imag(z)
   ans =
      3
```

The MATLAB answer "ans" is a variable that MATLAB creates to give the result of a calculation when we do not assign a variable to the results of `imag(z)` or `real(z)`.

3.1.1 Symbolic Variables

A symbolic variable is a variable that does not have a numeric value. We need to have installed the Symbolic Math Toolbox. This toolbox is necessary to define and use symbolic variables and functions. A symbolic variable can be defined with

```
a = sym('a')
x = sym('x')
```

or simply

```
syms a x
```

As with numeric variables, a symbolic variable starts with a letter. If a and x are symbolic variables, we can define symbolic functions. For example, we define the function $f(t)$ as

$$f(t) = 10 \sin(3\omega t + \theta) \tag{3.1}$$

In MATLAB we use:

```
>> syms w t theta
>> f = str2sym('10*sin(3*w*t + theta)')
   f =
      10*sin(3*w*t + theta)
```

To define a symbolic complex variable we can first define the real and imaginary parts as real variables and then define the new variable as complex. For example, if

$$z = x + iy \tag{3.2}$$

Then, the complex variable z is defined as

```
>>  syms x y real
>>  z = x + i*y
    z =
       x + y*1i
```

with x, y, z we now can perform calculations taking them as either real or complex.

We can also define complex variables with the instruction complex(a, b) where a and b are the real and imaginary parts, respectively. To obtain the complex conjugate of a complex number z we use the instruction conj(z)

```
>> conj(z) % returns the complex conjugate of z
   ans =
         x - y*1i

>>  conj(x)
```

Since x is real, conj(x) returns the same value. The product of a complex number z by its complex conjugate is its magnitude squared

```
>> mag_sq = z*conj(z)
   mag_sq =
        (x + y*1i)*(x - y*1i)
```

With the instruction **expand**, MATLAB rewrites the results as

```
>> magSq = expand(mag_sq)
   magSq =
       x∧2 + y∧2
```

And in a mathematical format with

```
>> pretty(magSq)
        2    2
    x  +   y
```

A variable that has been defined as complex can be made real with

```
>> syms x real
```

3.2 Functions

There exist two types of functions in MATLAB: those already predefined in MATLAB and those defined by the user. Some MATLAB functions fall within the set of elementary functions such as trigonometric, logarithmic, and hyperbolic functions among others. Some elementary functions were presented in Chapter 1.

As mentioned above, it is possible to have user-defined functions. We can define them in the **Command Window** or in a script file or in a function file in the MATLAB editor. To write a script or a function it is necessary to open the MATLAB editor and click on the **New** icon and select either **Script** or **Function**. We write the instructions in the editor window open. When finished, we save the file, name it with an appropriately, and run it with the **Run** icon or with the F5 key, or from the **Command Window** entering the file name followed by the key **Enter**. All files corresponding to scripts and functions are saved with the extension **.m**, and therefore, they are called **m-files**. We now show two examples, in the first one we enter all the instructions in the **Command Window** and for the second one we write a script.

Example 3.1 Calculation of the areas and perimeters of a hexagon and a triangle
We wish to calculate the area and perimeter for a hexagon and a triangle. The triangle is equilateral with side 8. The hexagon side is 10. The equation to calculate the area of a triangle with base L is

$$A_T = \frac{Lh}{2} \tag{3.3}$$

where h is the triangle's height. The height h is given by $h = L\sqrt{3}/2$. For the hexagon, the area is given by

$$A_H = \frac{3\sqrt{3}}{2}L^2 \tag{3.4}$$

and the perimeter is

$$P_H = 6L \tag{3.5}$$

where L is the side of the hexagon. In the Command Window we enter the following instructions:

```
>> Lt = 8; Lh = 10; %Triangle side = 8, hexagon side = 10.
>> h = sqrt(3)/2; % triangle's height.
>> % Now we enter the equations for the calculations needed.
>> Area_triangle = Lt*h/2 % circle's area.
   Area_triangle =
       3.4641
>> Area_hexagon = 3*sqrt(3)*Lh/2 % Hexagon's area.
   Area_hexagon =
       25.9808
>> Perimeter_hexagon = 6*Lh
   Perimeter_hexagon =
       60
```

In the previous example we have placed comments after the instructions. Comments can thus start at the end of the instruction or at the beginning of the row.

Example 3.2 Parabolic motion
A player throws a ball. The ball trajectory follows a parabolic path and this path makes an angle with respect to the horizontal. The initial velocity is v. We want to find the total distance traveled by the ball before it falls to the ground and the time the ball is in the air. The velocity vector components are $v_{ox} = |v| \cos \theta$

$$v_{ox} = |v| \cos \theta$$

$$v_{oy} = |v| \sin \theta$$

It can be shown that the time t the ball is in the air, the distance d traveled, and the ball path are given by

$$t = \frac{2v_{oy}}{g}$$

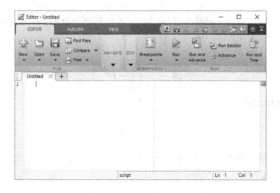

FIGURE 3.1: MATLAB Editor window.

$$d = v_{ox}t$$

$$y = \frac{v_{oy}}{v_{ox}}x - \frac{1}{2}\frac{g}{v_{ox}}x^2$$

Here g is the acceleration due to gravity. The data is $\theta = 40°$, $|v| = 45$ m/s. We can write a script to perform the desired calculations. The first thing to do is to click on the New m-file icon in the left top corner of the MATLAB window. This opens the MATLAB editor with an empty window to write our data, equations, and comments. The editor window is shown in Figure 3.1.

The instructions to do the calculations are written in the following m-file:

```
% This is file Example_3_2.m.
theta = 40; % initial angle in degrees.
theta_rad = theta*pi/180; % degree to radians conversion.
speed = 45; % initial speed.
g = 9.8; % acceleration due to gravity.
v0x = speed*cos(theta_rad);
v0y = speed*sin(theta_rad);
time = 2*v0y/g;
distance = v0x*time;
fprintf('Time elapsed %12.5f \n', time)
fprintf('Distance traveled by the ball %12.5f \n', distance)
% Calculations to plot the path.
x = 0:0.1:distance;
y = (v0y/v0x)*x - (1/2)*(g/v0x^2)*x.^2;
plot(x, y)
xlabel('Distance')
```

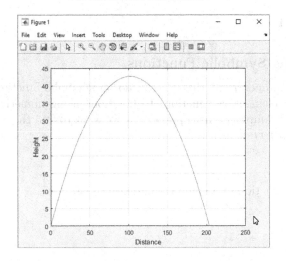

FIGURE 3.2: Ball's path.

```
ylabel('Height')
grid on
```

We save the script and we run it to obtain the following results:

```
Time elapsed 5.90315
Distance traveled by the ball 203.49344
```

The path's plot is shown in Figure 3.2.

3.2.1 MATLAB Elementary Functions

The elementary functions are grouped in the set Elementary Functions, `elfun`. The list can be displayed with

```
>>  help elfun
```

In this way we obtain a list of all the elementary functions in this set. There we can find all of the trigonometric, hyperbolic, logarithmic, exponential, complex, and other for rounding, residue, and sign functions.

Each one of these functions can be used with only typing the name with the argument between parentheses. For example, for the square root of 2 and the sine of π we use

```
>> sqrt(2)
```

```
>> sin(pi)
```

3.2.2   Using Symbolic Functions

There are some operations that we can do on symbolic functions. Thus, we can apply the elementary functions to functions using numeric or symbolic variables. For example, let us suppose that we want to find the first derivative of the function $f(x)$ defined by

$$f(x) = x^3 + 2x \qquad (3.6)$$

To define the function in a symbolic manner we can use

```
>> syms x
>>  f = x∧3 + 2*x
    f =
       x∧3 + 2*x
```

To evaluate a symbolic function in a given variable value we use the instruction

```
subs(f, a, value_of_a)
```

For example, for the function defined above, if we want to find its value when x = 2 we use

```
>> Value_of_f = subs( f, x, 2)
   Value_of_f =
      12
```

Now if we want to find the derivative we use the instruction `diff(f, n)` where n indicates what order derivative we want to find. For the function f we have then,

```
>> derivative_f = diff(f)
   derivative_f =
      3*x∧2 + 2
>> pretty(derivative_f)
         2
      3 x + 2
```

To integrate we use the instruction `int(f)`. Thus for f we have,

```
>> integral_f = int(f)
   integral_f =
      (x∧2*(x∧2 + 4) )
```

```
>> pretty(integral_f)
     2   2
   x ( x + 4)
   - - - - - - - - -
         4
```

And the definite integral from 1 to 2 can be found with

```
>> int(f, 1, 2)
   ans =
      27/4
```

3.2.3   Plots for Symbolic Functions

The instruction `plot` is useful to plot numeric vectors as we saw in Chapter 2. In the case of symbolic functions, the instruction we have to use is `ezplot`. For example, to plot the function $f(x) = x^2 \sin(x)$, we use

```
>> syms x
>> f = x∧2*sin(x)
>> ezplot(f)
>> grid on
```

and we obtain the plot shown in Figure 3.3. The figure includes a title with the function expression. We can also add limits to this instruction. If we want to plot from x = -20 to x = 20 we use

```
>> ezplot(f, -20, 20)
```

The resulting plot is shown in Figure 3.4. There we have added a legend with: `legend('Graph with ezplot')`.

Another instruction we can use is `fplot`. The same format applies to both `ezplot` and `fplot`. Thus, for the plot of the function

$$e^{-x/2}\sin^2(x)$$

in the range from -5 to +6 (in `fplot` we have to add the range) we can use:

```
>> fplot('exp(-x/2)*sin(x)∧2', [-2, 4])
>> legend('Plot with fplot')
>> grid on
```

This set of instructions produces the plot shown in Figure 3.5. Note the difference among the functions `plot`, `ezplot`, and `fplot`. The function `plot` must have the data in vector form while `ezplot` and `fplot` work on a symbolic

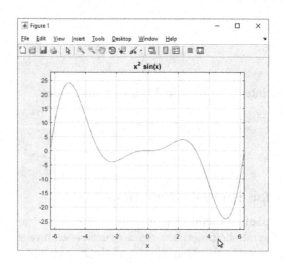

FIGURE 3.3: Plot of f(x) = x∧2*sin(x).

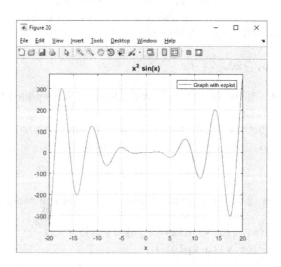

FIGURE 3.4: Plot of f(x) = x∧2*sin(x) in the range -20 to 20.

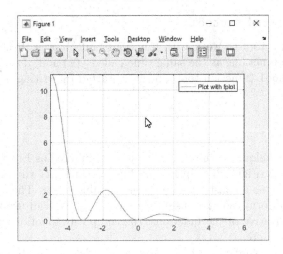

FIGURE 3.5: Plot of `f(x) = exp(-x/2)*sin(x∧2)` in the interval `[-5, 6]`.

function. That is, we have to generate the numeric data for `plot` in the form of two vectors, one vector for the independent variable and another one for the dependent variable.

3.2.4 Function Evaluation Using `eval` and `feval`

To evaluate a symbolic variable or function we can use the instruction `eval`. For example, to evaluate the function `sin(x)` at $x = \pi/2$, that we know it has unity value, we can use

```
>> eval ('sin(pi/2)')
   ans =
      1
```

The symbolic function must be between single quotes in the same way strings are treated besides being a valid MATLAB expression.

Another useful evaluation instruction is `feval`. This is used to evaluate symbolic expressions at a given number of points. For example, to evaluate the function `tan(x)` at the points `x = 0, 1, 2`. We do this first by defining x as `x = [0, 1, 2]` and then `feval('tan', x)`. Thus,

```
>> x = [1, 2, 3]
>> feval('tan', x)
   ans =
      1.5574 -2.1850 -0.1425
```

3.2.5   The Tool `funtool`

MATLAB has a tool to work with functions and observe their plots. It can work with two functions at the same time called $f(x)$ and $g(x)$. This tool has the name `funtool` (function tool) and it is started by writing in the Command Window

>> funtool

Three windows belong to `funtool`. They are numbered as Figures 1, 2 and 3 and are shown in Figure 3.6. In Figure 3.6c we can enter the functions `f(x)` and `g(x)` as well as a constant `a` and a set of values for `x`. This window has a keyboard to perform calculations on the functions. We can use the keyboard to find the derivative of the functions, the composite function `f[g(x)]`, the product of them, etc.

   After starting `funtool`, Figure 3.6c has by default the functions `f(x)` = `x`, `g(x)` = `1`, but these functions can be changed by users. Also the constant value is `a` = `1/2` and the range of `x` is `[ -2, 2]`.

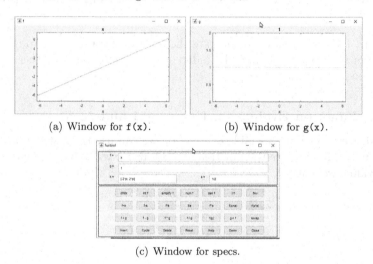

(a) Window for `f(x)`.   (b) Window for `g(x)`.

(c) Window for specs.

FIGURE 3.6: Windows for the function `funtool`.

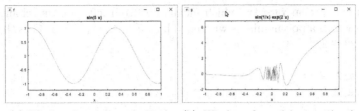

(a) Window for f(x) = sin(5*x). (b) Window for g(x) = exp(a*x) *sin(1/x).

(c) Window for specs.

FIGURE 3.7: Windows for the function funtool with a = 2 and [-1, 1].

3.3 Polynomials

Polynomials are a very special class of functions. A polynomial is a function of the form

$$p(x) = a_n x^n + a_{n-1} x^{n-1} + \ldots + a_1 x + a_0 \qquad (3.7)$$

The quantities a_k are called the polynomial coefficients. In MATLAB, a polynomial is represented as a row vector and the coefficients are given in descending order. For example, for the following polynomials we have

$x + 1$	is	[1 1]
$x - 1$	is	[1 -1]
$4x^2 + 2x - 3$	is	[4 2 -3]
$-2x^2 + \sqrt{7}x + 5$	is	[-2 $\sqrt{7}$ 5]

For the last example, we have

```
>> a = [-2 sqrt(7) 5]
   a
      = -2.0000 2.6458 5.0000
```

The instruction `fliplr`, which means flip from left to right, changes the order in which a polynomial is written. Thus, if `a_1 = [1 2 3 4]` is the polynomial $x^3 + 2x^2 + 3x + 4$, then

```
>> a_2 = fliplr(a_1)
   a_2 =
      4 3 2 1
```

represents the polynomial $4x^3 + 3x^2 + 2x + 1$. To evaluate a polynomial at a given value of x we can use the instruction `polyval(a, x)`. For the polynomial given by `poly` above, its value at $x = 9$ is

```
>> polyval(a, 9)
   ans =
      -133.1882
```

We can change the single valued x for a vector of x values as in

```
>> x = [1 -1 2 5 6 9];
>> polyval(a, x)
  ans =
       5.6458 0.3542 2.2915 -31.7712 -51.1255 -133.1882
```

To plot a polynomial we simply define the vectors x, y as in

```
>> x = linspace(-2, 2, 200); % Definition of the vector x.
>> a = [6 3 -7 0.4]; % Definition of the polynomial.
>> y = polyval(a, x); % Definition of the vector y.
>> plot(x, y) % Plotting of y vs. x.
>> grid on
```

The resulting plot is shown in Figure 3.8. To find the roots of a polynomial we have the instruction `roots`. For example,

```
>> zeros = roots(a)
   zeros =
      -1.3803
      0.8215
      0.0588
```

Since it is a third degree polynomial, three roots can be obtained which are called zeros. Note that the roots are in a column vector while the coefficients are in a row vector. If we have a set of roots, we can obtain the original polynomial with the instruction `poly`. For the polynomial given we have

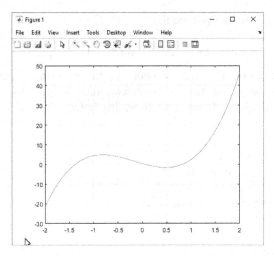

FIGURE 3.8: Plot of polynomial $6x^3 + 3x^2 - 7x + 0.4$.

```
>> a = poly(zeros)
a =
   1.0000 0.5000 -1.1667 0.0667
```

Note that the leading coefficient is unity. Multiplying this result by 6 we get the original polynomial as

```
>> a = 6*a
   a =
      6.0000 3.0000 -7.0000 0.4000
```

Note that both polynomials have the same roots. It may be possible in some cases to have a small imaginary part. This is due to round off errors and it should be neglected. This happens because MATLAB works with complex numbers. We can use the instruction **real** to eliminate any imaginary part.

The product of two polynomials is another polynomial. We can use the instruction **conv** to multiply two polynomials as in

```
>> conv (polynomial_1, polynomial_2)
```

For example:

```
>> poly_2 = [3 -4 7];
>> poly_1 = [4 -2 0 1];
>> conv(poly_1, poly_2)
   ans =
      12 -22 36 -11 -4 7
```

which corresponds to the polynomial

$$p(x) = 12x^5 - 22x^4 + 36x^3 - 11^2 - 4x + 7 \tag{3.8}$$

With `conv` we can only multiply two polynomials at a time. If we need to multiply three polynomials we first multiply two and then the result is multiplied with the remaining one. If we want to multiply $(x^2 - x - 1)(x^3 - 2)(x^2 - 3)$ we can use

```
>> poly_1 = [1 -1 -1];
>> poly_2 = [1 0 0 -2];
>> poly_3 = [1 0 -3];
>> pl1 = conv(poly_1, poly_2);
>> poly_final = conv(pl1, poly_3)

    poly_final =
        1 -1 -4 1 5 8 -6 -6
```

This can also be calculated with

```
>> pm = conv(poly_3, conv(poly_1, poly_2))

    pm =
    1 -1 -4 1 5 8 -6 -6
```

Both results give the polynomial

$$p_m(x) = x^7 - x^6 - 4x^5 + x^4 + 5x^3 + 8x^2 - 6x - 6$$

To divide polynomials we use the instruction `deconv` that has the format

$$[q, r] = deconv (a,b)$$

The polynomials q and r are the quotient and the residue of the division, respectively. For example, for the polynomials a and b we have

```
>> a = [1 6 24 50 77 84 64];
>> b = [1 4 9 16];
>> [q, r] = deconv(a, b)

    q =
        1 2 7 -12
    r =
        0 0 0 0 30 80 256
```

That is, the quotient and residue polynomials are

$$q(x) = x^3 + 2x^2 + 7x - 12$$

and

$$r(x) = 30x^2 + 80x + 256$$

To add polynomials they must have the same degree. If they do not have the same degree then we have to add zeros to the lower degree polynomial. For the polynomials a and b given above we have to add zeros to polynomial b as in

```
>> a = [1 6 24 50 77 84 64];
>> b = [0 0 0 1 4 9 16];
>> c = a + b

c =
    1 6 24 51 81 93 80
```

Then the polynomial c(x) is

$$c(x) = x^6 + 6x^5 + 24x^4 + 51x^3 + 81x^2 + 93x + 80$$

The derivative of a polynomial can be found with the instruction

```
polyder(polynomial)
```

For the polynomial a we have

```
>> polyder(a)

ans =
    6 30 96 150 154 84
```

A rational function is the quotient of two polynomials. That is, it is of the form

$$\frac{\text{polynomial } a}{\text{polynomial } b}$$

A rational function can be expanded in partial fractions. This is an expansion of the rational function to an expression of the form

$$\frac{p(x)}{q(x)} = \sum \frac{k_i}{x + p_i} + \frac{k_0}{x} + k_\infty x \tag{3.9}$$

In this expansion, p_i is called the i-th pole and k_i is the residue at pole p_i, k_∞ is the residue due to a pole at infinity, and k_0 is the residue for a pole at the origin. For example, for the polynomials $p(x)$ and $q(x)$ given by

$$p(x) = 10x + 20$$

$$q(x) = x^3 + 8x^2 + 19x + 12$$

The partial fraction expansion is given by

$$\frac{10(x+2)}{x^3 + 8x^2 + 19x + 12} = \frac{10(x+2)}{(x+4)(x+3)(x+1)} = \frac{-6.67}{(x+4)} + \frac{5}{(x+3)} + \frac{1.667}{(x+1)}$$

The poles of the rational function are p_i = -1, -3, -4, the residues at these poles are k_i= -6.667, 5, 1.667. Since the rational function does not have either a pole at infinity or a pole at the origin, $k_\infty = k_0 = 0$.

In MATLAB we can calculate the poles and residues with the instruction

```
[residues,poles,k_inf] = residue(poly_numerator,poly_denominator)
```

The results of this instruction are a vector with the residues, a vector of poles and the residue for the pole at infinity. For example,

```
>> num = [10 20];
>> den = [1 8 19 12];
>> [residues, poles, k] = residue(num, den)

    residues =
      -6.6667
       5.0000
       1.6667

    poles =
      -4.0000
      -3.0000
      -1.0000

    k =
      [ ]
```

The same instruction is used for the inverse operation. That is, if we give the poles, residues, and constant term we can obtain the polynomials of the rational function with the instruction residue as

```
>> [numerator, denominator] = residue ( residues, poles, k)
```

If we use the results of the previous example we obtain

```
>> residues = [1 2 3];
>> poles = [-1 + j, -1-j, -2];
>> k_inf = 2;
>> [n, d] = residue(residues, poles, k_inf)

    n =
       2.0000 14.0000 27.0000 - 1.0000i 20.0000 - 2.0000i

    d =
       1 4 6 4
```

The numerator and denominator polynomials are then

$$n(x) = 2x^3 + 14x^2 + (27 - i)x + (20 - 2i)$$
$$d(x) = x^3 + 4x^2 + 6x + 4$$

3.4 Curve Fitting

In many scientific and engineering applications, it is necessary to describe experimental data in an analytical form by means of a function. Thus, we can describe the experiment by a function which in many cases it is a polynomial. MATLAB allows us to fit data to a polynomial by using the instruction polyfit which has the format

```
p = polyfit(x, y, n)
```

where x, y are data vectors and n is the order of the polynomial p. For example, let us assume that we collected the data from an experiment in Table 3.1.

TABLE 3.1: Experimental data

Coordenadas	Valores
0	0
1	1
3	31
4	35
5	45
6	0

The plot is shown in Figure 3.10. We have changed the x-axis limits from -1 to 7 using from the main menu Edit → Axis Properties. The symbol '*k' indicates that the points are plotted with asterisks. We now use the instruction `polyfit` to find third and fifth degree polynomial approximations to the data points. First we obtain the third degree polynomial as:

```
>> x = [ 0, 1, 2, 3, 4, 5, 6];
>> y = [0, 1, 3.3, 2.2, 5.6, 4.4, 0];
>> p = polyfit( x, y, 3)

    p =
        -0.1583 1.0024 -0.3060 0.2190
```

That is, the polynomial we get is

$$p(x) = -0.1583x^3 + 1.0024x - 0.3060x + 0.2190$$

We now try a fifth order polynomial to see that the higher the order the better the approximation. Then,

```
>> p5 = polyfit(x, y, 5)

    P5 =
        -0.0087 0.0718 -0.1301 -0.1960 1.6957 -0.0618
```

and we now plot the data points and the third and fifth degree polynomials with:

```
>> p5v = polyval(p5, x1);
>> plot(x, y, '*r', x1, p1, x1, p5v, '-.')
>> legend('Data points', '3er order', '5th order')
```

The plot showing data points, the third order polynomial, and the fifth order polynomial are shown in Figure 3.9. We can readily see that the fifth order polynomial is a better approximation to the data points.

3.4.1   Cubic Spline Fitting

A cubic spline provides another way to curve fitting. A cubic spline is a set of third degree polynomials that exactly passes through the data points and in addition its first derivative is continuous at the data points. The instruction to obtain a cubic spline is

```
z = spline (data_points_x, data_points_y, x)
```

where `data_points_x` and `data_points_y` are the coordinates of the data points, and x is a vector of points where we wish the spline z to be evaluated.

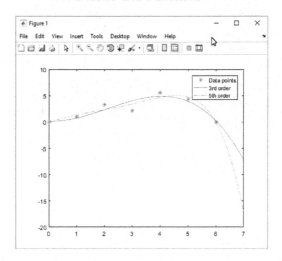

FIGURE 3.9: Plot of data points and third and fifth order polynomials.

For example, for data points `input_x`, `input_y` we want to find the spline; in addition, we wish the spline evaluated at the points given by the vector x defined below

```
>> input_x = [1 2 3 4 5 6];
>> input_y = [1 0 4.4 0 5.5 0];
>> x = linspace(1, 6, 100);
>> z = spline(input_x, input_y, x);
>> plot(input_x, input_y,'*r', x, z)
```

The variable z contains the spline data evaluated at the points in vector x. The plot of the data points and the spline is shown in Figure 3.10. We readily see that the spline provides an exact approximation at the data points. This is due to the fact that the spline is in fact a set of third degree polynomials.

3.4.2 The Tool Basic Fitting

We can also do a curve fitting using the MATLAB tool called `Basic Fitting`. This tool is available with the `Curve Fitting` Toolbox. It can be called from any plot available. For example, if we have sets of data points x, y we just plot them. For example, we plot the data points with an asterisk:

```
>> x = [1 2 3 4 5 6];
>> y = [1 0 4.4 0 5.5 0];
>> plot(x, y, '*')
```

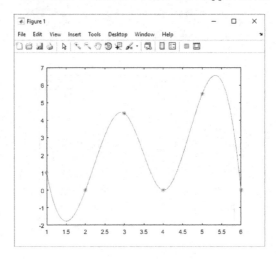

FIGURE 3.10: Plot of data points and the spline.

In the figures Tools menu we select the option for **Basic Fitting** as shown in Figure 3.11. This option opens the window of Figure 3.12 which gives the option to select the type of approximation. We can choose from the different degrees of polynomial approximation or spline approximation. As mentioned above, the best type of approximation is a cubic spline, but this option does not give an analytical expression as does the polynomial approximation. We select a 5th order polynomial approximation. We also select to plot the residuals, that is, the error between the points and the approximation. Also, the arrow at the bottom allows the display of a second and a third window to show the polynomial equation, if chosen in the first window. In the third window we can evaluate the approximation in a given set of points as shown in Figure 3.13. The original points, the approximation polynomial, and the evaluation at the points selected are shown in Figure 3.14.

3.5   Solution of Equations

To solve equations we can use the instruction **solve**. The argument for this instruction can be an equation such as

$$ax + b = 0 \tag{3.10}$$

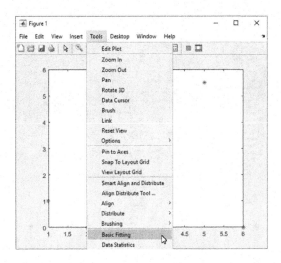

FIGURE 3.11: Choice of Basic Fitting from the Tools menu.

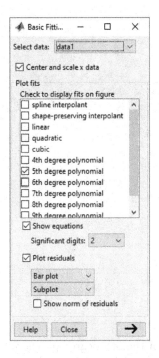

FIGURE 3.12: Window to select the interpolating polynomials.

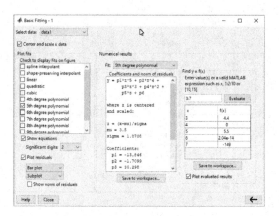

FIGURE 3.13: The three windows for the `Basic Fitting Tool`.

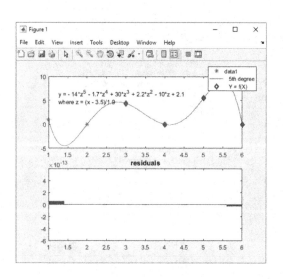

FIGURE 3.14: Curves of the 5th degree polynomial and spline in the top plot. Residuals plot in the bottom plot.

Or simply an expression as

$$ax + b \tag{3.11}$$

In this case, MATLAB equates to zero the expression. The equation must be written with a double equal sign, for example,

```
>> solve('a*x + b == 0')
   ans =
      -b/a

>> solve('a*x + b', 'x')
   ans =
      -b/a
```

Note that in both cases we get the same result. Also, a message may display indicating conditions on the values of a and b. We also see that even though in the first case we did not indicate what is the variable, MATLAB takes as unknowns the last few letters of the alphabet. Thus, the unkown is taken as x. If the unknown is either a or b we have to indicate this explicitly. If for example the unknown is a, we indicate this as

```
>> solve('a*x + b', 'a')
   ans =
      -b/x
```

In some cases we get a symbolic result which looks as a numeric one as in

```
>> f = solve('cos(x) = sin(x)')
   f =
      1/4*pi
```

We get a numeric value with

```
>> double(f)
   ans =
      0.7854
```

Or with

```
>> eval(f)
   ans =
      0.7854
```

If MATLAB cannot find a symbolic solution, then it gives a solution that looks like a number as in

```
>> x = solve('exp(x) = tan(x)')
   x =
        -226.19467105846511316931032359612
```

This is not a numeric value, rather it is a symbolic one. To evaluate this symbolic result we use `eval` as in

```
>> eval(x)
   ans =
        -226.1947
```

We can also solve equations with the instruction `fzero`. This function uses a function defined in an m-file. The format for the instruction `fzero` as

```
x_sol = fzero(function, x_ini)
```

where `x_sol` is a vector with the solutions of the equation `function = 0`. `x_ini` is an initial value for the solution. For example, to find the solutions of

$$(x - 3)x = 7 \tag{3.12}$$

We have that the equation is

$$x_{sol} = (x - 3)x - 7 \tag{3.13}$$

Using the MATLAB editor we write the following m-file and save it,

```
function x_sol = fx(x);
% This is file fx.m
x_sol = (x - 3).*x - 7;
```

Giving the initial value as x = 4, we now run the instruction

```
>> fzero('fx', 4)
   ans =
        4.54138126514911
```

If we change the initial condition we get another solution as

```
>> fzero('fx', -1)
      ans =
        -1.54138126514911
```

A way to find initial values is to plot the function and from there make an

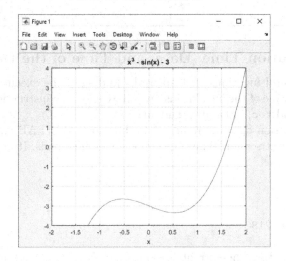

FIGURE 3.15: Plot of the function x∧3 - sin(x) - 3.

educated guess about the initial values by observing the zero crossings. We can plot the functions with the instructions **fplot** and **ezplot** as in

```
>> ezplot( 'function')
>> fplot('function')
```

For example, to find the roots of

$$f(x) = x^3 - \sin(x) - 3$$

We use

```
>> syms x
>> ezplot(x∧3 - sin(x) - 3)
>> grid on
>> axis([-2 2 -4 4])
```

This gives the plot of Figure 3.15. We see that there is a zero crossing close to x = 2. We use this as an initial condition. Then,

```
>> root = fzero('x∧3 - sin(x) - 3', 2)
     root =
         1.5874
```

3.6   Execution Time, Date, and Time of the Day

There are several functions that allow the finding of the execution time, the date, and the time of the day. In this section we cover the instructions cputime, clock, tic and toc, date, datenum and now.

The instruction cputime indicates for how long MATLAB has been working in the current session. The response is given in seconds. If we execute this instruction we get

```
>> cputime
   ans =
       106.2188
```

which indicates that the current session started 106.2188 seconds ago.

We can use cputime to find the time taken to execute a set of instructions. For example,

```
>>  t_initial = cputime;
>> x = linspace (0, 0.1 , 1.00);
>> a = sin(log10(x));
>> t_final = cputime - t_initial
   t_final =
       0.0781
```

Thus, we see that the time elapsed in the above calculations was 0.0781 seconds.

The instructions tic and toc are used to measure the time elapsed in the execution of a set of instructions. For example,

```
>> tic;
>> x = linspace (0,0.01,100);
>> x = tan (x).* sin (x);
>> toc;
       Elapsed time is 0.054738 seconds.
```

The time elapsed depends upon the processor speed and of the number of programs running at the same time.

With the instruction clock we obtain a row vector with the date and time. The data format is: year, month, day, hour, minutes, second.

```
>> clock
   ans =
       1.0e+003 *
```

2.0180 0.0020 0.0260 0.0100 0.0380 0.0007

which indicates that the year is 2018, the second month, that is **February**,
the 26th day, and the time of the day is 10 **hr**, 1 **minute**, and 7 **seconds**.

The instruction **date** gives the date in a string,

```
>> a = date
   a =
      26-Nov-2018
```

As with any other string, we can write only a few characters of the chain

```
>> a(3:7)
   ans =
      -Nov-
```

MATLAB counts the days beginning on January 1st, 0000. This is only
a reference point. The instruction **datenum** displays the code number for any
date in particular. For example, the date January 14, 2010 gives:

```
>> n = datenum(2018, 01, 14)
   n =
      737014
```

If we wish to know how many days have passed between two dates we can use
datenum. For example, to know how many days have passed between October
5, 2011 and October 5, 2022 we use

```
>> n = datenum('10-5-2020')- datenum('10-5-2019')
   n =
      366
```

And we see that 2020 is a leap year because the result is 366 days. The format
for the date can be one of several ones. For example

```
>> datenum('January-14-2025')
>> datenum('january-14-2025')
>> datenum('14-jan-2025')
>> datenum('1/14/25')
>> datenum('14-jan-2025')
   ans =
      739631
```

All of them give the same result. The hour can also be included inside **datenum**,

```
>> format bank
>> datenum('january-14-2025 12:00 PM')
    ans =
        739631.50
```

The decimal part is due to the fact that 12 PM is the half of the day.

The instruction **now** gives the date in a numeric format. For the day **Feb 26, 2018** we get

```
>> now
    ans =
        737117.45
```

The decimal part is because we are entering this instruction before noon which is before the half of the day.

3.7   Concluding Remarks

We have presented the way MATLAB works with variables and functions, either user-defined or MATLAB predefined functions. Even though MATLAB has more than ten thousand predefined functions, in most cases it will not have a particular function needed by a user. Thus, it is of paramount importance that users know how to define their own functions.

Chapter 4

Matrices and Linear Algebra

4.1   Introduction

As mentioned in Chapter 1, MATLAB was originally designed to carry out operations with matrices, and received the name of **Matrix Lab**oratory that subsequently was reduced to MATLAB. Thus, it deserves special attention to learn the way in which MATLAB works with them.

Matrices are practically found in all the areas of knowledge. In this manner, it is possible to find applications of matrices not only in the mathematics areas, but also in a great deal of applications in engineering, in physics, in finance, in accounting, in the arts, music, in anthropology, chemistry and biology, to mention a few.

Another important application of matrices is given in the solution of systems of simultaneous equations because the equations' coefficients are arranged in a matrix. Simultaneous equations systems appear in many problems in engineering such as in optimization of systems, automatic control, food engineering, chemical engineering, etc. Chapters 9 to 14 present several matrix applications in the solution of problems.

A matrix is defined as an array of numbers (or entities). From this definition of a matrix, we also refer to a matrix as an array. Both terms are used indistinctly.

MATLAB has an array editor that allows users to introduce the elements of a matrix or to modify the elements of a matrix given previously and already stored as a MATLAB variable. This array editor will be described later in the chapter.

Additionally to the basic operations with matrices and vectors, MATLAB has the potential to handle arrays by means of indexing. In this manner carrying out calculations using arrays becomes very efficient and fast.

Finally, the chapter covers cell and structure arrays. Cell arrays are also matrices, but the fundamental difference is that they can handle combinations of different variable types such as names of people or things and numbers. These arrays allow us to handle data in a more efficient and more practical way. A structure is similar to a cell but each element is identified by a name.

The chapter begins with a description of matrices and the basic operations that can be performed with them. It continues with vectors and how to compute dot and vector products. A section is dedicated to matrix and vector functions. A section is presented to solve simultaneous equations systems, and for LU factorization of a matrix. The chapter ends with cells arrays and structures.

4.2   Matrices

MATLAB handles in matrix form all the variables defined in a MATLAB session, whether they are user-defined or they are in the predefined functions inside MATLAB. It is then convenient to consider the concept of a matrix that any book of linear algebra defines as:

A Matrix is an array of numbers or objects.

In this way, a matrix can be an array of numbers, letters, objects, or any thing. The only condition that a matrix should satisfy to be a matrix is for it to be an ordered array. For example, a table of numbers is a matrix. The pixels of an image also form a matrix. These two matrices are ordered in a plane. If an image changes with time, as is the case of television images or motion pictures, then we are adding another dimension to the definition of such matrices and thus we are talking about multidimensional matrices. The objects that form the matrix are called **elements** of the matrix.

An example of a matrix is

$$A = \begin{bmatrix} 1 & -3 & 0 \\ 8 & -6 & 4 \\ 2 & 7 & 5 \\ -4 & 9 & 5 \end{bmatrix}$$

which has 4 rows and 3 columns. We say that matrix A is of dimension 4×3. If a matrix has n rows and m columns, it is a $n \times m$ matrix. The quantity $n \times m$ is called the dimension of the matrix. In MATLAB, the matrix A is defined by

$$A = [\,1 \quad -3 \quad 0\,;8 \quad -6 \quad 4\,;2 \quad 7 \quad 5\,;-4 \quad 9 \quad 5\,]$$

That is, we have separated the elements of the same row by a space (they can be also separated with a comma) and the rows are separated with a semicolon. Matrix A is shown in Figure 4.1. Each element from the matrix A has a position i, j which corresponds to its position in the ith row and in the jth column. The numbers i and j are called the indices of the (i, j) element. For example, in matrix A the element 8 has the position 2, 1 while the element 7 has the position 3, 2. The command

$$A(i, j)$$

defines the element of row i and column j. If the number of rows is equal to the number of columns, that is, if n = m we say that matrix A is a square matrix of dimension n. Other commands are given in Table 4.1.

TABLE 4.1: Matrix commands

Matrix command	Description
A(a:b, c:d)	Submatrix consisting of rows a to b and columns c to d.
A(k, :)	Vector formed by the elements in the k-th row.
A(:, j)	Vector formed by the elements in the j-th column.

MATLAB has an array editor. To illustrate its use we consider matrix A given above. We write again matrix A in the Command Window. This is shown in Figure 4.2 where we have also displayed the Workspace window. In this window we see matrix A and in the Workspace we see the icon for the variable A specified as a 4×3 array. If we double click on this icon the Array Editor will open as a new window attached to MATLAB as shown in Figure 4.2. There we can clearly see the element values which now can be either modified or deleted. We can even add new elements to matrix A. In this example we add a fifth row to A. The elements that are going to be added are

5   10   -7

FIGURE 4.1: MATLAB main window.

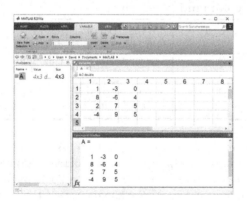

FIGURE 4.2: MATLAB array editor window.

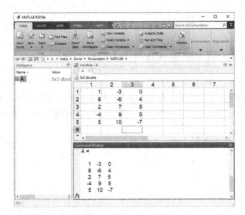

FIGURE 4.3: MATLAB and array editor with new elements.

TABLE 4.2: Special matrices

Matrix name	Description
ones(m, n)	Matrix of m×n dimension where each element has unity value.
ones(n)	Square matrix of order n where each element has unity value.
zeros(m, n)	Matrix with every element having zero value.
zeros(n)	Square matrix of order n where each element has zero value.
eye(n)	Square identity matrix of order n
eye(m, n)	Matrix of m×n dimension with 1's in the main diagonal. The remaining elements are zeros.

We see in Figure 4.3 the elements added. If in the `Command Window` we write A, the matrix will be displayed as shown in Figure 4.3. Another way to open the array editor is by writing

>> `open name_of_the_matrix`

The array editor is then open showing the elements of the variable selected. Some special matrices are shown in Table 4.2.

4.3   Basic Operations with Matrices

The basic operations with matrices are addition, subtraction, multiplication, and division. The addition and subtraction of two matrices can be carried out if both matrices are the same size n × m. For example for A, B and C given by

$$A = \begin{bmatrix} 1 & -3 \\ 9 & 5 \\ 7 & -6 \end{bmatrix}, \quad B = \begin{bmatrix} 4 & -2 & 0 \\ 3 & 6 & 1 \end{bmatrix}, \quad C = \begin{bmatrix} 2 & -2 \\ 6 & -1 \\ 5 & -3 \end{bmatrix}$$

Matrix A has dimension 3×2, matrix B dimension is 2×3 and for matrix C the dimension is 3×2. For addition or subtraction, it can only be done for A and C because they have the same size 3×2. Then:

```
>>  A + C
    ans =
          3          -5
         15           4
         12          -9

>> A − C
    ans =
         -1          -1
          3           6
          2          -3

>>  C − A
    ans =
          1           1
         -3          -6
         -2           3
```

The multiplication of two matrices can be carried out if the number of columns of the first matrix is equal to the number of rows of the second matrix. In our matrices, the product can be performed for A and B, C and B, as well as that of B and C, B and A. We show the result of A*B

```
>> D = A*B
    D =
         -5         -20         -3
         51          12          5
         10         -50         -6
```

If one of the matrices is a scalar, that is, it is a matrix of dimension 1×1, the computations can also be done. In this case, in the case of addition, the scalar is added to each element of the matrix. For multiplication, each element of the matrix is multiplied by the scalar.

The inverse of a square matrix A is a matrix M such that

$$AM = MA = I$$

where I is the identity matrix (the `eye(n)` matrix in MATLAB). The inverse of a square matrix A is obtained with

```
>> inv(A)
```

For the division of square matrices two possibilities exist, namely A/B and A\B. The result of A/B is the same as A*inv (B) while A\B is the same as inv(A)*B. For the matrices A and B given by

```
A = [1    2      3; 4    5      -6; 9    8     7]

B = [4    5     -7; -6    3     2; 0    1     -1]
```

we have that

```
>> A/B
   ans =
         4.3750          2.7500          -28.1250
         1.7500          0.5000           -5.2500
        16.8750          9.7500         -105.6250

>> B\A
   ans =
        -2.2667         -3.4833           4.4500
         1.7333          3.7667          -4.3000
         0.9333          0.3167          -0.9500
```

For powers of matrices, there exist two possibilities. If p is a scalar A∧p gives the matrix A to the scalar p. On the other hand p∧A gives the scalar p to the A power. Some examples illustrate this for p = 2

```
>> A∧p
   ans =
         24          55
         44         101

>> p∧A
   ans =
```

```
1.0e+003 *
0.4455        1.0202
0.8162        1.8738
```

4.3.1   The Dot Operator

There is another operator to work with arrays and matrices. It is called the **dot operator** and it refers to operations term-to-term between two matrices of the same dimension. This operator is useful to multiply, divide, and to obtain powers of arrays. The operations are term-by-term. The complete operator is formed by placing a dot before the operator. The operations for matrices A and B of the same dimension produce a new matrix C whose elements are given by:

For multiplication:

$$C = A.*B$$

where the elements of C are given by:

$$c_{ij} = a_{ij}*b_{ij}$$

For division:

$$C = A./B$$

where the elements of C are given by:

$$c_{ij} = a_{ij}/b_{ij}$$

and for exponentiation:

$$C = A.\wedge B$$

where the elements of C are given by:

$$c_{ij} = a_{ij}\wedge \$textttb_{ij}$$

Other operations with matrices are given in Table 4.3.

4.4   The Characteristic Polynomial

For every square matrix we associate a polynomial called the characteristic polynomial. The instruction `poly` produces the characteristic polynomial $p(x)$ of a square matrix A which is defined as

TABLE 4.3: Matrix operations

Matrix operation	Description
diag (A)	Vector with the elements of the main diagonal.
inv A	Inverse of A.
A'	Transpose of A.
transpose (A)	Transpose of A.
det (A)	Determinant of A (if A is a square matrix).
rank(A)	Rank of A.
trace (A)	Sum of the elements of the main diagonal of A.
norm A	Norm of A.
A∧c	Matrix A to the cth power.
A.∧c	Each element of A to the cth power.
A/B	Same as A*inv(B).
A\B	Same as inv(A)*B.
A.*B	Term-to-term product.
A./B	Term-to-term division.
poly(A)	Finds the characteristic polynomial of A.

$$p(x) = det(A - xI)$$

where I is the identity matrix and det calculates the determinant of the matrix
A - xI. For example, for the matrix A given by

```
>> A = [1 2 3; 5 6 7; 9 10 11 ]

    A =
        1         2         3
        5         6         7
        9        10        11

>> poly(A)
    ans =
        1.0000      -18.0000      -24.0000       0.0000
```

which means that the characteristic polynomial of the square matrix A is

$$p(x) = x^3 - 18 \ x^2 - 24 \ x$$

4.5 Vectors

A vector is a matrix with a single row called row vector, or with a single
column, called column vector. Vectors obey matrix rules. As an example, the

row vector given by:

$$x = [1 \ 3 \ -7 \ 4]$$

is a row vector of dimension 4, while

$$y = [\ 9\ ;\ 18\ ;\ -5\ ;\ 6\ ;\ -7\ ;\ 2\ ;\ 4\ ;\ 3\ ;\ 11\ ;\ 17\ ]$$

is a column vector of dimension 10.

Note that a vector has the same structure in MATLAB as a polynomial does, according to the definition of Chapter 3 for polynomials. Nevertheless, the operations that are carried out on them are completely different.

If we only desire to display on the Command Window some elements of a vector we use indexing. For example y(k: m) specifies that we only wish to display from the to k^{th} to the mth elements. For example

```
>> y(2:4)
    ans =
      18
      -5
       6
```

and y(i: j: k) gives the elements from the ith to the kth, but separated by j units. Then,

```
>> y(2 : 3 : 10)
    ans =
      18
      -7
       3
```

The index j can be also negative. For example:

```
>> y(9 : -3 : 3)
    ans =
      11
       2
      -5
```

Some operations with vectors are given in Table 4.4. In this table we have that a and b are n-dimensional vectors and c is a scalar. The operations preceded by a dot are term-to-term operations. We now present some examples.

If vectors a and b and the scalar c are given by

TABLE 4.4: Vector operations

Operation	Result
a + c	$a_1+c, a_2+c, \ldots, a_n+c$
a*c	$a_1{}^*c, a_2{}^*c, \ldots, a_n{}^*c$
a + b	$a_1 + b_1, a_2 + b_2, \ldots, a_n + b_n$
a.*b	$a_1 * b_1, a_2 * b_2, \ldots a_n * b_n$
a./b	$a_1/b_1, a_2/b_2, \ldots a_n/b_n$
a.\b	$b_1/a_1, b_2/a_2, \ldots, b_n/a_n$
a.∧c	$a_1 \wedge c, a_2 \wedge c, \ldots, a_n \wedge c$
c.∧a	$c \wedge a_1, c \wedge a_2, \ldots, c \wedge a_n$
a.∧b	$a_1 \wedge b_1, a_2 \wedge b_2, \ldots, a_n \wedge b_n$

```
a = [ -3, 7, 1 ],    b = [ 2, 4, 3 ],    c = 2
```

Then, we can perform the following operations:

```
>> a = [ -3, 7, 1 ],    b = [ 2, 4, 3 ],    c = 2
>> a + c
   ans =
        -1    9    3

>> a*c
   ans =
        -6   14    2

>> a.∧b
   ans =
         9  2401    1

>> a.*b
   ans =
        -6   28    3

>> a.\b
   ans =
     -0.6667    0.5714    3.0000
```

The product of two vectors is only possible if we observe matrix multiplication rules. In this way, it is possible to multiply two vectors if the first one is a row vector with m columns and the second one is a column vector with m rows, or if we wish to multiply a column vector of n rows by a row vector of n columns. For example, if we have the vectors

```
t = [a b c];
r = [w x y z];
s = [1; 2; 3; 4; 5];
u = [10; 20; 30];
```

We will be able to perform only the following operations:

```
>> syms a b c w x y z
>> t = [a b c]; % Row vector.
>> r = [w x y z]; % Row vector.
>> s = [1; 2; 3; 4; 5]; % Column vector.
>> u = [10; 20; 30]; % Column vector.

>> t*u
   ans =
        10*a + 20*b + 30*c % This is the dot product.

>> u*t
   ans =
        [ 10*a, 10*b, 10*c]
        [ 20*a, 20*b, 20*c]
        [ 30*a, 30*b, 30*c]
```

Any other operation that we perform will indicate an error because the dimensions of the vectors do not allow it.

In a similar way to the case of matrices, the basic operations that involve scalars can also be done. In this way, if we have a vector v and we wish to subtract 2 to each element we do it with: v - 2.

If we have a matrix A of dimension n×m and a vector b of dimension m, (that is, the number of columns of A is equal to the number of rows of b) we can multiply A*b. If A and b are

$$A = \begin{bmatrix} 3 & -2 & 0 \\ 4 & 9 & 17 \\ 2 & -4 & 9 \\ 6 & 2 & -5 \end{bmatrix}, \quad b = \begin{bmatrix} 3 \\ 2 \\ -7 \end{bmatrix}$$

Then A*b is

```
>> A = [3 -2 0; 4 9 17; 2 -4 9; 6 2 -5];
>> b = [3; 2; -7];
>> A*b
   ans =
         5
       -89
       -65
```

57

4.5.1 Norm of a Vector

The norm of a vector is a generalization of the length of a vector. The p-norm of an n-dimensional vector is defined by

$$||x||_p = (x_1^p + x_2^p + ... + x_n^p)^{1/p}$$

We use:

```
>> norm( x, p)
```

to obtain in MATLAB the norm of a vector. The default value for p is 2, corresponding to the Euclidean norm also called 2-norm. The 1-norm is the sum of the absolute values of the vector components. That is:

$$||x||_2 = \sqrt{x_1^2 + x_2^2 + ... + x_n^2}$$

$$||x||_1 = |x_1| + |x_2| + ... + |x_n|$$

The ∞-norm, also called Chebyshev norm or the maximum norm is given by the greatest magnitude of the vector's components. Thus, for the vector x = [3 4 -2]', if we wish to find the 1-norm:

```
>> norm( x, 1)
    ans =
       9
```

For the Euclidean norm norm(x, 2) or simply norm(x)

```
>> norm(x, 2)
    ans =
       5.3852
```

```
>> norm(x)
    ans =
       5.3852
```

The Euclidean norm is the vector magnitude given by

$$|x| = ||x||_2 = \sqrt{x_1^2 + x_2^2 + ... + x_n^2}$$

And for the ∞-norm

```
>> norm(x, Inf)
   ans =
      4
```

4.5.2   Vector Generation

A special type of vector is the one that we generate to evaluate a function in a given interval. For example, if it is desired to evaluate a function f(x) in an interval [a, b]. If the distance between consecutive points is linear, we can use:

$$x = \text{linspace}(a, b, n)$$

This instruction generates a vector of n elements equally spaced between a and b. For example,

```
>> x = linspace(1, 10, 10)
   x =
      1 2 3 4 5 6 7 8 9 10
```

To create a similar vector we can use

$$x = a : \text{increment} : b;$$

where a is the initial point and the following points are:

```
a + increment, a + 2*increment,..., a + k*increment,.., b
```

Depending on the value of the increment, the final value can be different from b. For example,

```
>> x = 3 : 0.7 : 6
   x =
      3.0000 3.7000 4.4000 5.1000 5.8000
```

To create a logarithmic spaced vector we use

$$x = \text{logspace} (a , b , n)$$

where the first point is 10^a, the last point is 10^b and there are n points in the interval. For a = 2, b = 5 and n = 4 we have

```
>> x = logspace(2, 5, 4)
   x =
      100 1000 10000 100000
```

4.6 Dot and Cross Products

There are two important products involving vectors. They are the **dot product** and the **cross product** also known as **vector product**. The dot product result is a scalar and the cross product one is a vector.

4.6.1 Dot Product

The dot product of two vectors is also called the scalar product and inner product. The dot product result is a scalar. For vectors x and y, the dot product is given by:

$$(x, y) = x_1{}^*y_1 + x_2{}^*y_2 + \ldots + x_n{}^*y_n$$

MATLAB finds the dot product with:

```
>> dot ( a , b )
```

For vectors a and b given below, we have

```
>> a = [ 2 ; 3 ; -2 ; 1 ];
>> b = [ 3 ; -8 ; 7 ; 4 ];
>> c = dot (a , b)
   c =
       -28
```

The dot product can also be found with:

$$(a, b) = |a| * |b| \cos(\theta)$$

where θ is the angle between both vectors. We can readily see that if the vectors are orthogonal, that is, the angle between vectors a and b is $\theta = 90°$ and we have $\cos(90°) = 0$, and the dot product is equal to zero.

The dot product can also be evaluated if we transpose the first vector to make it a row vector and then we multiply it by the second column vector. That is,

```
a'*b
```

For the previously defined vectors we have

```
>> a'*b
   ans =
       -28
```

4.6.2   Cross Product

The cross product a × b, also called vector product, is another vector orthogonal to the plane formed by vectors a and b. The cross product can only be evaluated for three-dimensional vectors. MATLAB finds the cross product with

```
>> cross(a, b)
```

For vectors a and b given below, we find the cross product with

```
>> a1 = [ 9 ; 6 ; 23 ];
>> b1 = [ -27 ; 11 ; 12 ];
>> cross(a1, b1)
   ans =
      -181
      -729
       261
```

which corresponds to vector [-181, -729, 261].

4.7   Matrix and Vector Functions

It is possible to evaluate functions of matrices and vectors. When we request to evaluate a function of a matrix, MATLAB makes the necessary calculations (transparent to the user) to produce the result. Evaluating matrix functions is not an easy task since it requires the evaluation of eigenvalues and eigenvectors, and in some cases generalized eigenvectors. This is required to transform the matrix into the required form involving a modal matrix and a diagonal one, or possibly a quasidiagonal one. To show how easy it is to evaluate matrix and vector functions in MATLAB, let us consider the row vector given by x = [2 3 -7]. To find the function sin(x) we simply write in the Command Window:

```
>> x = [ 2 3 -7 ]
>> y = sin(x)
   y =
        0.9093 0.1411 -0.6570
```

which is another vector whose components are the sine of each component of the vector x.

Functions defined in MATLAB, as well as user defined ones, can be used with vectors and matrices. Thus, for any function $f(x)$, if A is a matrix, $f(A)$ is given by

$$f(A) = \begin{bmatrix} f(a_{11}) & f(a_{12}) & \ldots & f(a_{1n}) \\ f(a_{21}) & f(a_{22}) & \ldots & f(a_{2n}) \\ \vdots & & & \\ f(a_{n1}) & f(a_{n2}) & \ldots & f(a_{nn}) \end{bmatrix}$$

And if v is a vector, then:

$$f(v) = [\ f(v_1)\ \ f(v_2)\ \ \ldots\ \ f(v_n)]$$

where v can be either a row vector or a column one. For example, for a polynomial defined by

$$p(x) = x \wedge 3 + 2 * x \wedge 2 - 7 * x + 17$$

where x can be a scalar, a vector, or a matrix. If D is a matrix given by

$$D = \begin{bmatrix} 2 & 5 \\ 4 & 9 \end{bmatrix}$$

then

```
>> D = [2 5 ; -4 9 ];
>> D∧3 + 2*D∧2 - 7*D + 17
   ans =
       -281  507
       -375  405
```

4.8 Systems of Simultaneous Linear Equations

For a set of simultaneous linear equations such as

$$a\,x + b\,y = c$$

$$d\,x + e\,y = f$$

MATLAB can solve them, symbolically, with the instruction solve in the following way (note the double equal sign):

```
>> syms a b c d e f x y
>> [x, y] = solve( [a*x + b*y == c, d*x + e*y == f ],[x, y])
   x =
       -( b*f - c*e )/(a*e - b*d)
```

```
    y =
       (a*f - c*d)/(a*e - b*d)
```

The instruction `linsolve` solves the same set of equations, but we have to write the system in matrix form. Thus, for the set given above, it can be solved with

```
>> syms a b c d e f x y
>> A = [a b; d e];
>> v = [c; f];
>> x = linsolve(A, v)
    x =
       [ -(-e*c + f*b)/(a*e - b*d)]
       [ (a*f - c*d)/(a*e - b*d)]
```

A set of simultaneous equations can also be solved to give a numerical solution by taking the inverse of the matrix A. Then, the vector of unknowns is given by

$$x = \text{inv}(A)*b$$

For example, for the system of equations given by:

$$A = \begin{bmatrix} 2 & -2 & 0 & 0 \\ -2 & 6 & -2 & 0 \\ 0 & -2 & 6 & 2 \\ 0 & 0 & -2 & 0 \end{bmatrix}, \quad b = \begin{bmatrix} 5 \\ 0 \\ 0 \\ 0 \end{bmatrix}$$

To solve this system of equations, we first evaluate the inverse of A and then we multiply it by b:

```
>> A = [2 -2 0 0;-2 6 -2 0;0 -2 6 -2;0 0 -2 0];
>> b = [5; 0; 0; 0];
>> x = inv(A)*b
    x =
       3.7500
       1.2500
            0
      -1.2500
```

The same result can be obtained by the instruction

```
>> x = A\b
    x =
       3.7500
       1.2500
```

```
        0
    -1.2500
```

which gives the same result.

4.8.1   LU Factorization

LU factorization allows us to write a square non-singular matrix A as the product of a lower triangular matrix L and an upper triangular matrix U, that is, we can write A as

$$A = L\ U$$

LU factorization can be performed with the instruction lu(A) and the instruction format is

```
>>  [ L , U] = lu (A)
```

For the matrix A given by

```
>>  A = [1 2 3 4; 5 6 7 8; 9 10 11 12; -3 4 -6 9]
    A =
          1    2    3    4
          5    6    7    8
          9   10   11   12
         -3    4   -6    9

>>  [L, U] = lu(A)
    L =
         0.1111    0.1212    1.0000         0
         0.5556    0.0606    0.5000    1.0000
         1.0000         0         0         0
        -0.3333    1.0000         0         0

    U =
         9.0000   10.0000   11.0000   12.0000
              0    7.3333   -2.3333   13.0000
              0         0    2.0606    1.0909
              0         0         0         0
```

If matrix A is the matrix of a system of linear equations in the form Ax = b, then the solution is

```
x = U\(L\b)
```

If b = [1 ; 2; -1; 6]   then for the matrix A given above we have

```
>>  x = U \(L\b)

    x =
            1.0e+15*
        -5.2984
         6.9938
         1.9074
        -3.6029
```

4.9   Eigenvalues and Eigenvectors

A very important topic in matrix algebra concerns the way to find eigenvectors and eigenvalues. Eigenvectors are unique vectors that for a square matrix A of dimension n and a vector x of size n, satisfy the following equation:

$$Ax = \lambda x$$

That is, when we multiply A by x we obtain the same vector but multiplied by a constant λ called the **eigenvalue** of A corresponding to the **eigenvector** x. In other words, vector x only changes magnitude but keeps the same direction. Eigenvalues can be either real or complex numbers. To find eigenvectors and eigenvalues we use the instruction

$$[V, D] = eig(A)$$

The instruction eig returns two matrices. Matrix V has as columns the eigenvectors of A. Matrix D is a diagonal matrix whose elements of the main diagonal are the eigenvalues. As an example, let us consider matrix A given by

```
>>  A = [4 0 1 0; 2 2 3 0; -1 0 2 0; 4 0 1 2]
```

Then, eigenvectors and eigenvalues can be found with

```
>>  [V, D] = eig(A)
```

and the results are:

```
    V =
            0        0      0.2887      0.2887
          1.0        0     -0.2887     -0.2887
```

```
       0      0     -0.2887    -0.2887
       0     1.0     0.8660     0.8660
```

D =
```
    2   0   0   0
    0   2   0   0
    0   0   3   0
    0   0   0   3
```

Thus, we see that the eigenvalues are: $\lambda_1 = 2$, $\lambda_2 = 2$, $\lambda_3 = 3$, and $\lambda_4 = 3$, and the corresponding eigenvectors are the columns of the matrix V.

Some matrices happen to have repeated eigenvalues and we cannot find all eigenvectors to be linearly independent. The procedure then is to find generalized eigenvectors. For a detailed treatment on generalized eigenvectors, readers can see reference 2. The procedure to find generalized eigenvectors using MATLAB is

$$[V, J] = \text{jordan } (A)$$

where V is a matrix whose columns are the generalized eigenvectors and J is an almost diagonal matrix. The elements in the main diagonal are the eigenvalues and the elements in superdiagonal are equal to 1. We say that matrix J is in Jordan canonical form. For example, for matrix A given above

```
>> A = [2 2 1 1; 0 2 -1 0; 0 0 2 0; 0 0 0 1]
   A =
      2 2  1 1
      0 2 -1 0
      0 0  2 0
      0 0  0 1

>> [V, D] = jordan(A)
   V =
      -1.0000 1.0000  -1.0000   1.0000
            0      0   0.5000  -0.2500
            0      0        0  -0.5000
       1.0000      0        0        0

   D =
      1 0 0 0
      0 2 1 0
      0 0 2 1
      0 0 0 2
```

We can see that in the superdiagonal we have 1's only above the eigenvalues with multiplicity greater than unity.

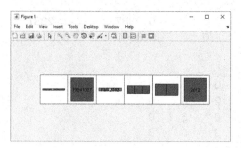

FIGURE 4.4: Cell array.

4.10 Cell Arrays

Cell arrays are matrices where we can store variables of different type in each cell. Each cell can be identified in the same way that matrix elements are identified. To show how a cell array can be created and used, let us store information about students in a cell array called `Student`. Each student will be assigned a column in the cell array. The information stored is: `Name`, `ID` number, `Semester`, `Homework` grades, `Exam` grades, `Year`. This is done in the following way:

```
Student{1, 1} = 'Ophelia Cervantes';
Student{1, 2} = 19941027;
Student{1, 3} = 'Fall 2012';
Student{1, 4} = [ 10 9.8 8 ];
Student{1, 5} = [ 9.9 8.9 ];
Student{1, 6} = 2012;
```

This technique to address arrays is called Content Addressing.
To add a second student to the cell array we simply write

```
Student(2, 1) = {'Laura Michele'};
Student(2, 2) = {19890510};
Student(2, 3) = {'Fall 2012'};
Student(2, 4) = {[10 10 7.8]};
Student(2, 5) = {[10 9.6]};
Student(2, 6) = {2013};
```

Note that in this case we enclosed the data with braces. This manner to enter information is known as cell indexing. We can see a cell by typing:

```
Student{2, :}
```

```
ans =
   Laura Michele
ans =
    19891005
ans =
   Fall 2019
ans =
    10.0000 10.0000 7.8000
ans =
    10.0000 9.6000
ans =
    2013
```

We can "see" the array elements with

```
cellplot(Student)
```

to obtain Figure 4.4.

4.11 Structures

Structures are arrays that can also store different types of information, but in this case each cell is identified by a field, that is, a name. Let us consider the cell arrays of the previous section. This information can be stored in a structure in the following way:

```
Student.Name = 'Gary Baez';
Student.ID = 19941027;
Student.Semester = 'Fall 2012';
Student.Homework = [ 10 9.8 8 ];
Student.Exams = [ 9.9 8.9 ];
Student.Graduation = 2012;
```

In this structure we are defining six fields and each field is making reference to a different subject. All these fields are grouped under the variable Student. If we write this variable in the Command Window we obtain the information stored there, as in:

```
>> Student
```

If we only wish to see a specific field we only write Student and the field name separated by a dot. For example if we wish to see the data for the field Semester we write

>> Student.Semester

To change or to add data we proceed as in the case of matrices. Thus, if we wish to change the Homework data to 10 9.9 9.8, then we only write

>> Student.Homework = [10 9.9 9.8];

To add a third exam we concatenate the new data to the previous data, as in

>> Student.Exams = [Student.Exams, 9.7];

To add a second student we write:

>> Student(2).Name = 'Laura Michele'

The variable Student now has two elements. The first element is Student(1) and the second element is Student(2). We can add data to the new Student as in

```
Student(2).ID = 19890510;
Student(2).Semester = 'Fall 2012';
Student(2).Homework = [10 10 7.8];
Student(2).Exams = [10 9.6];
Student(2).Graduation = 2013;
```

We can add now as many students as we need by following the same procedure.

4.12 Concluding Remarks

In this chapter we have covered the main aspect of the different arrays we can have in MATLAB. We started with matrices and vectors, the operations that we can work with them. Many of these operations and operators are unique to MATLAB. In addition, we covered eigenvectors and eigenvalues as well as Jordan canonical matrices. Finally, the topics of cell arrays and structures are described.

Chapter 5

Calculus

5.1 Introduction

MATLAB allows users to perform a number of operations on functions, such as differentiation and integration. We can also find limits and check for continuity at a point. These operations can be done in a numerical way or in a symbolic one. In addition, we can solve differential equations with great simplicity in the steps involved in the process. Such simplicity is an indication of how powerful is MATLAB to solve complicated tasks. In the case of operations involving symbolic functions, the result is also a symbolic function. However, the resulting function can be evaluated numerically or plotted to see its behavior. The availability of symbolic computations is due to the existence of the Symbolic Toolbox available from The Mathworks, Inc. This toolbox has to be installed to be able to perform symbolic computations.

The chapter is organized as follows: We start with limits of functions and sequences. We continue with function differentiation and integration. Finally, we show how to solve differential equations, both numerically and symbolically.

5.2   Limits of Functions

Limit calculation for sequences and functions is an important topic in calculus. The limit A of a function $f(x)$ as x approaches x_o is given by

$$A = \lim_{x \to x_0} f(x) \qquad (5.1)$$

We say that A is the limit and it is unique. MATLAB provides the instruction limit to find the limit of a function. The format is

limit(f, x, x0)

For example, for the function $f(x) = x^2$, the limit as x approaches 3 can be found with

```
>> syms x
>> limit( x∧2, x, 3)
   ans =
      9
```

And the limit of $g(x) = \sqrt{x^2 + 1}$ as x approaches $infty$ is found with

```
>> syms x
>> limit(sqrt(x∧2 + 1), x, Inf )
   ans =
      Inf
```

In the definition of limit there is no restriction as to how x should approach x_0. However, in some cases we need to specify the way x approaches x_0. For real variables, x can approach x_0 from the left or from the right. These limits are called lateral limits. We can find the limit from the right as

limit(f, x, a, 'right')

and for the limit from the left the instruction is

limit(f, x, a, 'left')

For example, if the function $f(x)$ is

$$f(x) = \frac{|x - 3|}{x - 3}$$

If we evaluate the left and the right limits we obtain

```
>> syms x
>> f = abs(x - 3)/(x - 3);
>> limit(f, x, 3, 'left')

    ans =
      -1

>> limit(f, x, 3, 'right')

    ans =
      1
```

If the limit of a function exists, it is **unique**. Thus, in the case of this function the limit as x approaches 3 does not exist.

The derivative of a function $f(x)$ at a point $x = x_0$ is defined as a limit. The definition of derivative is:

$$\left.\frac{df(x)}{dx}\right|_{x=x_0} = \lim_{x \to x_0} \frac{f(x) - f(x_0)}{x - x_0} \qquad (5.2)$$

For the function $f(x)$ defined by

$$f(x) = \frac{1}{x - 2}$$

the derivative is

$$\frac{df(x)}{dx} = -\frac{1}{(x - 2)^2}$$

Using limits, we can find the derivative at the point $x = x_o$ as

```
>> syms x0
>> limit((1/(x - 2)-1/(x0 - 2))/(x - x0), x, x0)
    ans =
      -1/(x0 - 2)^2
```

which is the expected result.

Limits can be nested. Nested limits appear frequently when we have functions of several variables. For example, in the function

$$f(x,y) = 2xy - x^2$$

We can find the limits as x approaches 2 and as y approaches 3 as

```
>> syms x y;
```

```
>> f = 2*x*y - x∧2;
>> limit(limit(f, x, 3), y, 2)
   ans =
      3
```

We can nest as many limits as we wish but it is recommended not to nest more than two limits.

5.3   Limits of Sequences

A sequence is a function whose domain is the set of positive integers. We usually write a sequence as $a = \{a_n\}$ and it is understood that the value of n goes from 1 to ∞. We say that a sequence converges to a number A as n approaches ∞ if for each $\epsilon > 0$ there exists a positive integer N such that for all $n > N$ we have $|a_n - A| < \epsilon$. The real number A is called the limit of $\{a_n\}$ as n approaches ∞. If a sequence does not converge we say that it diverges.

We can also find limits of sequences $\{a_n\}$ with MATLAB. Let us consider the sequence

$$a_n = \sqrt[n]{\frac{1+n}{n^2}}$$

It is desired to find

$$\lim_{n\to\infty} a_n = \lim_{n\to\infty} \sqrt[n]{\frac{1+n}{n^2}}$$

In MATLAB we use

```
>> syms n
>> limit(((1 + n)/n∧2)∧(1/n), Inf)
   ans =
      1
```

As another example, let us consider the sequence

$$a_n = \frac{n^2 + n}{n^2 - n}$$

whose limit we can find with

```
>> an = (n∧2 + n)/(n∧2 - n);
>> limit(an, n, Inf)
   ans =
```

1

Now, if the sequence is

$$a_n = \frac{3n^2 + 4n}{2n - 1}$$

We can look for the limit as

```
>> an = (3*n^2 + 4*n)/(2*n - 1);
>> limit(an, n, Inf)
      ans =
      Inf
```

which indicates that this sequence diverges.

5.4   Continuity

A function $f(x)$ is continuous at a point $x = a$ if

$$\lim_{x \to a} f(x) = f(a) \tag{5.3}$$

If the limit does not exist we say that the function is discontinuous at that point. If the limit exists but it is not equal to $f(a)$ it is said that the function has a discontinuity at $x = a$. Let us consider the functions

$$f(x) = \frac{\sin x}{x}, \; g(x) = \sin \frac{1}{x}$$

The limit of $f(x)$ as x approaches $x = a$ can be found with

```
>> syms x a
>> limit(sin(x)/x, x, a)
      ans =
      sin(a)/a
```

From this result we might think that the limit does not exist at $x = 0$. But when we find the limit as x approaches $x = 0$ we get

```
>> limit( sin(x)/x, x, 0)
      ans =
      1
```

For $x = a$

```
>> limit(sin(1/x), x, a)
```

```
ans =
    sin(1/a)
```

But for $x = 0$

```
>> limit(sin(1/x), x, 0 )

ans =
    NaN
```

which indicates that the limit is in the interval [-1, 1]. If we now evaluate the lateral limits we get

```
>> limit(sin( 1/x ), x, 0, 'left')
ans =
    NaN

>> limit ( sin ( 1/x ), x, 0, 'right')
ans =
    NaN
```

which indicates that the limit does not exist and, therefore, this function is discontinuous at $x = 0$.

Now, let us consider the function $f(x, y)$

$$f(x, y) = \frac{xy}{x^2 + y^2}$$

The domain of this function is the whole plane except for the point (0, 0). To find the limit as (x, y) approaches a point (x_0, y_0) in the plane, we can approach this point from any direction in the plane. For example, to find the limit we can use

```
>> syms x y
>> limit(limit( x*y/(x∧2 + y∧2), x, 0), y, 0)
ans =
    0

>> limit(limit(x*y/(x∧2 + y∧2), y, 0) , x, 0)
ans =
    0
```

indicating that this function is continuous at the point (0, 0). We can also use a set of straight lines passing through the origin. For example, the set of lines

$$y = mx$$

where m is the slope of the lines that pass through the origin. Using this equation, the function $f(x, y)$ becomes

$$f(x) = \frac{mx^2}{x^2 + mx}$$

Then the limit is

```
>> syms x m
>> limit((m*x^2 )/(x^2 + m*x ), x, 0)
   ans =
       0
```

We can use any family of curves. For example, for the family of parabolas

$$y^2 = mx$$

Then, $f(x, y)$ now becomes

$$f(x, y) = \frac{m^{1/2}x^{3/2}}{x^2 + mx}$$

and the limit is

```
>> limit(m^(1/2)*x^(3/2))/(x^2 + m*x), x, 0))
   ans =
       0
```

which again proves that the function is continuous at the origin.

5.5 Derivatives

Let a function $f(x)$ be defined at any point in the interval $[a, b]$. The derivative of $f(x)$ at $x = x_0$ is defined as

$$f'(x) = \lim_{x \to x_0} \frac{f(x) - f(x_0)}{x - x_0} \tag{5.4}$$

If the limit exists, the function $f(x)$ is called differentiable at the point $x = x_0$. If $f(x)$ is differentiable at a point, the function is continuous at that point.

MATLAB can evaluate derivatives in a symbolic manner using the instruction diff. To see how it works let us consider the function

$$f(x) = -3x^3$$

We know that its derivative is

$$f'(x) = -9x^2$$

Now, using MATLAB we get

```
>> syms x
>> f = -3*x∧3;
>> diff(f)
   ans =
      -9*x∧2
```

To find the second derivative of $f(x)$ we use

```
>> diff(f, 2)

   ans =

     -18*x
```

In the case of a function of several variables, we must indicate with respect to which variable we wish to find the derivative. For example, for the function $f(x) = 2x^3 y + 3x/y$, we wish to differentiate first with respect to y and then with respect to x. Then, we use

```
>> syms x y
>> f = 2*x∧3*y + 3*x/y;
>> f1 = diff(f, y)
   f1 =
     2*x∧3 - 3*x/y∧2

>> f2 = diff(f1, x)
   f2 =
     6*x∧2 - 3/y∧2
```

If we do not indicate the variable with respect to which we wish to differentiate, MATLAB differentiates with respect to the last few letters in the alphabet. Thus, in the case of the function $f(x) = axy + y$, MATLAB finds

```
>> f(x) = a*x*y + y;
>> f1 = diff(f)
   f1 =
     a*y
```

which is the derivative with respect to x.

```
>> f1 = diff(f, x)
   f1 =
     a*y
```

and this is the same result. For the derivative with respect to y:

```
>> f1 = diff(f, y)
   f1 =
       a*x + 1
```

If we differentiate with respect to a we have:

```
>> f1 = diff(f, a)
   f1 =
       x*y
```

Now, let us see the following results:

```
>> f = a*x*y + y;
>> f1 = diff(f)
   f1 =
       a*y
```

which is the derivative with respect to x. Now, the resulting function is independent of x. Then,

```
>> f1 = diff(f1)
   f1 =
     a
```

Since $f1$ only has as variables y and a, the derivative is taken with respect to y which belongs to the set of the last few letters in the alphabet. Now, the only variable is a and the derivative is taken with respect to a, as in

```
>> f1 = diff(f1)
   f1 =
       1
```

As another example, if $f(x) = x^n$, the derivative with respect to x is

```
>> syms x n
>> f = x∧ n;
>> diff (f)
     ans =
     x∧n*n/x
```

If we wish to differentiate with respect to n we must explicitly write it as

```
>> diff(f, n)
   ans =
   x∧n*log(x)
```

Note that in the case of a function of several variables $f(x, y)$, the derivative that we obtain with `diff` is the partial derivative

$$\text{diff}(f, \ x) = \frac{\partial f}{\partial x} \quad \text{and} \quad \text{diff}(f, \ y) = \frac{\partial f}{\partial y} \quad (5.5)$$

5.6   Integration

Integration is defined for continuous functions and for piecewise continuous functions. Thus, if a function is continuous in a region, then the integral of $f(x)$ is defined as another function $F(x)$ such that

$$F(x) = \int f(x)dx \quad (5.6)$$

The function $F(x)$ satisfies the condition

$$f(x) = \frac{dF(x)}{dx} \quad (5.7)$$

F is also called the antiderivative of $f(x)$.

MATLAB calculates the integral of a function $f(x)$ with the instruction `int(f)`. For example, for the function $f(x) = 2x$, the integral is given by

```
>> int(2*x)
   ans =
   x∧2
```

In the case of functions of several variables we have to indicate with respect to which variable we wish to integrate. For example, for the function $f(x) = 2x^2y$, its integral with respect to y is given by

```
>> int(2*x∧2*y, y)
   ans =
   x∧2*y∧2
```

The definite integral is evaluated by giving the limits after the function specification in the instruction `int`. Thus, for the function $2x^2 + 5x$, the integral between the limits 2 and 3 can be obtained with:

```
>> int(2*x∧2 + 5*x, 2, 3)
```

```
ans =
    151/6
```

Improper integrals such as

$$\int_0^\infty \frac{e^{-x}\sin x}{x}dx$$

can be evaluated in the same way as definite integrals. In this case we have

```
>> syms x
>> f = (exp(-x)*sin (x) )/x;
>> int(f, 0, Inf)
    ans =
        1/4*pi
```

Some integrals do not converge. For example, the function $f(x) = \tan(x)$ has a singular point at $x = \pi/2$, thus, to evaluate

$$\int_0^{\frac{\pi}{2}} \tan x dx$$

We use,

```
>> syms x b
>> f = tan (x);
>> int(f, x, 0, pi/2)
    ans =
        Inf
```

and then we conclude that the integral does not exist.

Sometimes, MATLAB is unable to find a closed form solution. In those cases we have to use a numerical solution. There are several instructions to integrate numerically. One of them is the instruction **quad**. It uses a Simpson rule to find the integral. The format is the same as in the symbolic instruction int. For example, for the function $f(x) = \sin(x)/x$, the integral

$$\int_0^{\frac{\pi}{2}} \sin x dx$$

can be found with

```
>> quad('sin(x)', pi/2, pi)
    ans =
        1.0000
```

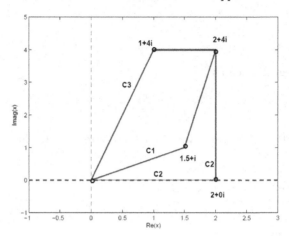

FIGURE 5.1: Integration paths.

The function $f(x)$ can be defined in an m-file

```
function y = myfunction(x)
y = sin(x);
```

and then,

```
>> quad(@ myfunction, pi/2, pi)
   ans =
   1.0000
```

With this instruction we can find line integrals. For example, the line integral

$$\int \frac{s+1}{s-1-2i} ds$$

We see that the function to be integrated has a singular point at $s = 1 + 2i$ and thus the path must not pass through this point. We wish to find the value of the integral from the point 0 to $2 + 4i$. In Figure 5.1 we see three possible paths. Because the function is analytic everywhere except at the singular point, the integral along any one of these paths would give the same value. For path C_1 we first integrate along the segment that starts at the origin and ends at $1.5 + i$ and then from the point $1.5 + i$ to the end point in the path $2 + 4i$. We do this with

```
>> quad('(s + 1)./(s−1−2*i )', 0, 1.5 + i)...
       + quad('(s + 1)./(s−1−2*i)', 1.5 + i, 2 + 4*i)
```

```
ans =
    -4.2832 + 10.2832i
```

We now integrate along C_2

```
>> quad( '(s+1)./(s−1−2*i)', 0, 2)...
            + quad('(s + 1)./(s−1−2*i)', 2, 2 + 4*i)


    ans =
        -4.2832 + 10.2832i
```

We obtain the same result if we integrate along path C_3.

Let us consider now the function $|s|$. We know this function is not analytic on the s-plane and, therefore, its integral depends upon the path we choose for the integral. On C_1 we obtain

```
>> quad( 'abs(s)', 0, 1.5 + i)...
            + quad('abs(s)', 1.5 + i, 2 + 4*i)


    ans =
        2.8911 + 10.1356i
```

On C_2 we get

```
>> quad('abs(s)', 0, 2) + quad('abs(s)', 2, 2 + 4i)


    ans =
        2.0000 + 11.8315i
```

and along C_3 we obtain

```
>> quad( 'abs(s)', 0, 1 + 4i)...
            + quad( 'abs(s)', 1 + 4i, 2 + 4i)


    ans =
        6.3421 + 8.2462i
```

As we see, the integral in the three cases yields a different value.

The instruction quad employs the Simpson method for the evaluation of the integral. Another method available for integral evaluation is the trapezoidal method which is available with the instruction trapz.

5.7   Series

Let us consider the sequence $\{a_n\}$ where the elements a_n can be numbers or functions. The formal sum of them

$$a_0 + a_1 + a_2 + ... + a_n + ... = \sum_{n=0}^{\infty} a_n$$

is called an infinite series. The partial sums are defined by

$$S_n = \sum_{k=0}^{n} a_k$$

The partial sums are a sequence. If the limit of the partial sums

$$\lim_{n \to \infty} S_n = S$$

exists, then S is called the sum or the limit of the series. In this case, we say that the series converges and

$$S = \sum_{n=0}^{\infty} a_k$$

If the limit does not exist we say that the series diverges.

MATLAB can find the sum of a series of the form

$$S = \sum_{n=a}^{b} f_n \tag{5.8}$$

with the instruction symsum that has the format

$$S = \text{symsum}(f, n, a, b)$$

where f_n is the nth term in the series, n is the index, and a and b are the lower and upper limits, respectively. For example for the power series

$$S = \sum_{n=0}^{\infty} x^n \tag{5.9}$$

We can find the sum with

```
>> syms x n
>> f = x^n;
>> sum = symsum ( f , n , 0 , inf )
    sum =
        piecewise([1 <= x, Inf], [abs(x) < 1, -1/(x - 1)])
```

which indicates that for $x >= 1$ the result is infinity, that is, the series diverges, and for $|x| < 1$ the series converges and the result is $1/(1-x)$.

We recall from calculus that this series converges if $|x| < 1$. Thus, although MATLAB has found a general solution for the sum, we must be careful about the conditions for convergence. We can check this by substituting values for x, say $x = 0.5$ and $x = 2$. We have then,

```
>> f = 0.5∧n;
>> sum = symsum ( f, n, 0, inf)
   sum =
      2

>> f = 2∧n;
>> sum = symsum ( f, n, 0, inf)
   sum =
      Inf
```

In the first case we see that the series converges, but for $x = 2$ the series diverges.

As another example, let us consider the series,

$$\sum_{n-0}^{\infty} \frac{2n+3}{(n+1)(n+2)}$$

Using MATLAB we can find the sum as

```
>> f = (-1)∧n*(2*n + 3)/(n + 1)/(n + 2);
>> sum = symsum ( f, n, 0, inf)
   sum =
      1
```

Thus, the series converges.

We know that the harmonic series diverges. This series is given by

$$a_n = \frac{1}{n}$$

We can see this with

```
>> an = 1/n;
>> sum = symsum ( an, n, 1, inf)
   sum =
      Inf
```

5.8 Differential Equations

A differential equation is an equation involving derivatives. They arise in many fields of physics, mathematics, engineering, economy, finance, etc. Examples of phenomena described by differential equations are Newton's second law of mechanics, Hooke's law, electric circuits, a simple pendulum, cash flow, and birth rate.

An nth order ordinary differential equation can be described by the equation

$$\frac{d^n y(t)}{dt^n} = f\left(y(t), \frac{dy(t)}{dt}, \frac{d^2 y(t)}{dt^2}, ..., \frac{d^{n-1} y(t)}{dt^{n-1}}, t\right) \qquad (5.10)$$

This equation can be solved by MATLAB. The representation of a derivative in MATLAB, for the symbolic solution of differential equations is Dy for the first derivative, D2y for the second derivative, and for the nth derivative

$$\frac{d^n y(t)}{dt^n} \rightarrow \texttt{Dny} \qquad (5.11)$$

Then, the following differential equation is described in MATLAB notation as

$$\frac{d^2 y(t)}{dt^2} = 1 + \frac{dy(t)}{dt} + y^2 t \rightarrow \texttt{D2y(t) = 1 + Dy(t) + y\textasciicircum2*t} \qquad (5.12)$$

A differential equation can be solved symbolically with the instruction dsolve. As an example, the differential equation

$$\frac{d^2 y(t)}{dt^2} + y(t) = 4$$

can be solved in MATLAB with

```
>> dsolve( 'D2y + y = 4')
    ans =
      4 + C1*sin(t) + C2*cos(t)
```

C1 and C2 are constants which can be found from the initial conditions. If the initial conditions are

$$y(0) = 1 \quad y'(0) = 0$$

Then we add them to the instruction dsolve as

```
>> dsolve ('D2y + y = 4', 'y (0) = 1, Dy(0) = 0')
   ans =
   4 - 3*cos(t)
```

Note that the independent variable is not explicitly defined, but MATLAB has as the default independent variable the letter *t*. We may, however, assign the independent variable. For example, if we want x as the independent variable, then we use

```
>> dsolve('D2y + y = 4' , 'y(0) = 1, Dy(0) = 0', 'x')
   ans =
   4 - 3*cos(x)
```

For the second order differential equation

$$\frac{d^2y(t)}{dt^2} + a\frac{dy(t)}{dt} + by(t) = c \tag{5.13}$$

The instruction dsolve is

```
>> y = dsolve('D2y + a*Dy + b*y = c')
   y =
   1/b*c + C1*exp(-1/2*(a + (a^2 - 4*b)^(1/2))*t) ...
      + C2*exp(-1/2*(a-(a^2-4*b)^(1/2))*t)
```

We use the instruction **pretty** to display the solution in a more readable format,

```
>> pretty(y)
```

```
        / /               2      \\       / /               2      \\
c       | |  a     sqrt(a -4b)| |       | |  a     sqrt(a -4b)| |
-+C1 exp| -t| -  - ----------- | | +C2 exp| -t| - + ---------- | |
b       \ \  2         2        //       \ \  2        2       //
```

For the system of linear differential equations

$$\frac{dx(t)}{dt} = 3x(t) + 4y(t)$$

$$\frac{dy(t)}{dt} = -7x(t) + 6y(t)$$

we can also use the instruction **dsolve** as

```
>> [x, y] = dsolve('Dx = 3*x + 4*y','Dy = -7*x + 6*y',...
        'x(0) = 2, 'y(0) = 1')
    x =
    -1/103*exp(9/2*t)*(-206*cos(1/2*t*103^(1/2))....
        25cm -2*103^(1/2)*sin(1/2*t*103^(1/2)))

    y =
    -1/103*exp(9/2*t)*(25*103^(1/2)*sin(1/2*t*103^(1/2))...
        -103*cos(1/2*t*103^(1/2)))
```

A technique for solving linear differential equations is by separation of variables. To see how we can use **dsolve** we consider the differential equation

$$y \sin(t) \ dt - (1 + y^2) dy = 0$$

that can be rewritten as

$$\sin(t) \ dt = \frac{1 + y^2}{y} dy$$

We now use **dsolve** on each term and then equate both results. For the left hand term we have

```
>> syms t
>> int(sin(t))
    ans =
        -cos(t)
```

And for the right term we have

```
>> syms y
>> int(( 1 + y^2)/y)
    ans =
        1/2*y^2 + log(y)
```

We now equate both results and use the instruction **solve** to find the solution of the differential equation,

```
>> solution = solve('-cos(t) = 1/2*y^2 + log(y)')
    solution =
    exp(-1/2*lambertw(exp(-2*cos(t)))-cos(t))
```

where **lambertw** is the Lambert W function. The Lambert W function is the inverse function of

$$f(W) = We^W$$

5.8.1 Numerical Solution of Differential Equations

MATLAB provides instructions to numerically solve a differential equation. Two of the most used ones are ode23 and ode45. ode23 uses Runge-Kutta second and third order methods. With ode23, a second order method is used first and then a third order method. A similar description applies for ode45 but in this case fourth and fifth order methods are used. The difference between the two methods is that ode23 uses more crude tolerances. Both methods are one step solvers. The format for these instructions is

```
[t, y] = ode23( F, [ t_initial, t_final ], Yo)
[t, y] = ode23( F, [ t_initial, t_final ], Yo)
```

Here, F is a string of text and it indicates where the differential equation is defined, usually an m-file. t_initial and t_final are the initial and final times for the simulation and Yo is the initial condition. As an example, let us consider the following differential equation

$$\frac{dy}{dx} = -2yt$$

We define this equation in an m-file dy.m and then we save it in the MATLAB directory. The m-file is

```
function yderivative = dy(t, y)
% This is file dy.m
yderivative = -2*y*t;
```

we now solve the differential equation with ode23 with

```
>> y0 = 2;
>> [ t , y ] = ode45('dy', [0, 2], y0);
>> plot(t, y)
>> grid
```

The result is plotted in Figure 5.2. The exact solution is $(y(t) = 2 * \exp(-t^2))$. If we plot this exact solution in the same plot we find that the match is exact. We can plot the exact solution with

```
>> t = 0:0.1:2;
>> y1 = 2*exp(-t.^2);
>> plot(t, y1)
```

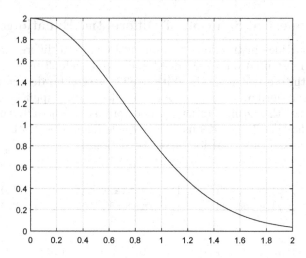

FIGURE 5.2: Solution of the differential equation $dy/dt = -2yt$.

5.9   Concluding Remarks

MATLAB can also work calculus. In this chapter we work with symbolic techniques to differentiate, integrate, find limits of sequences and series, and to solve differential equations. However, it can also perform numerical computations. We presented examples for line integral evaluation and numerical solution of differential equations.

Chapter 6

Programming in MATLAB

6.1 Introduction

MATLAB provides a very powerful high-level programming language, embedded in a computational environment, known as m-language. A very important advantage, when compared to other programming languages such as C,

FORTRAN, Java, and VisualBasic, to name a few, is that variables do not have to be defined at the beginning of the program, but rather they are defined automatically as they are written for the first time in the program. In addition, the type of the variables can be changed along the program depending upon the programmer's needs. Another very important advantage is that all the instructions and functions used and programmed in MATLAB can be used in the program. Thus, instructions and functions such as the ones used for optimization, image processing, integration, matrix inversion, to solve differential equations, etc., can be used in the program. This makes the m-language much more powerful than any other programming language. In this chapter we cover many of the instructions used to write programs using m-language and give some examples.

6.2   Creating m-files

Files containing programs using m-language are called m-files. They have an extension .m. There are two types of m-files: scripts and functions. Scripts are programs that neither have input data nor output data explicitly declared in the call to the script. On the other hand, functions do have input data and produce output data. The variables of a script or a function are called local variables to the program. Thus, in other programs, they are not available unless they are declared as global variables. To see the contents of a program we just need to enter the instruction `type` followed by the file name. For example, if there is a file containing the program `quadratic.m`, we do the following in the `Command Window`:

    type quadratic

An m-file containing a function must have the following parts:

1. Comments that start with the symbol %

2. Input data instructions

3. Instructions that execute the action for the function

4. Output instructions

5. It may have an end instruction to end the function

To create and edit an m-file we use the m-File editor/Debugger that comes with MATLAB. It is shown in Figure 6.1. We select `New Script`. This opens the editor for m-files where we write the script. Once we finish writing the m-file we save it in a directory. In this window we can run the file by pressing the `Run` icon or by typing the name of the function or script.

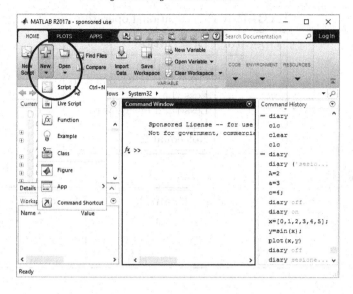

FIGURE 6.1: Menu to create an m-file.

6.3 Basic Programming Instructions in MATLAB

In this section we present the basic programming instructions that we may require in a MATLAB m-file. We show its use with simple, but useful programs where we can appreciate the powerful m-language.

6.3.1 The Instruction `if-end`

A very important characteristic in any programming language is the ability to change the order in which a set of instructions is executed. This usually depends on a set of conditions. If they are satisfied, the sequence has a certain order. If they are not, then the sequence changes the order in which instructions are executed. The set of conditions depends on variables in the program and/or input variables. MATLAB has instructions that do these kind of decisions. One of such instructions is the `if` statement. It has the general form:

```
if condition
    statements
end
```

TABLE 6.1:   Relational operators

Operator	Description
$>$	Greater than
$>=$	Greater than or equal to
$<$	Less than
$<=$	Less than or equal to
$==$	Equal to
$\sim=$	Not equal to

TABLE 6.2:   Logic operators

Operator	Description	Example	Precedence
$\sim$	NOT	$(a > b)$	1
&	AND	$(a > b)\&(x < 5)$	2
		OR	$(a > b) \mid (x < 5)$	3

Note that this `if` statement is similar to those in other programming languages. The indentation is not needed but it is useful for clarity. The condition is a logical expression whose value can be true or false. It can contain relational operators. Those are shown in Table 6.1. If the condition is satisfied and its value is true, then all the statements between the `if` and the `end` keywords are executed. Conditions are of the form

$$e_1 \, \mathcal{R} \, e_2$$

where e_1 and e_2 are arithmetic expressions and $\mathcal{R}$ is one of the relational operators in Table 6.1. Sometimes we need more that one relational operator or condition. In that case they are linked by a logical operator. The logical operators are listed in Table 6.2. A small example will show the use of the instruction `if`.

Example 6.1 Use of the instruction if
In a school, a passing grade has to be greater than 7.5. An m-file reads the input data and indicates if the student passes or fails the course. This file is created with the MATLAB editor and after writing it and before executing it we save it in the directory Chapter_6 with the name `six_1.m`. The m-file is:

```
% File six_1.m
% Reads in a grade and writes out
% if the student passes or fails.
%
% Reads in the grade
calif = input ('Give me the grade:\n');
if calif >= 7.5
```

```
      fprintf ('Congratulations, you passed.')
end
if calif < 7.5
      fprintf('I am sorry. You failed.')
end
```

Now we run the program with a grade of 5 and a grade of 9,

```
>> six_1
Give me the grade:
5
I am sorry. You failed.
>> six_1
Give me the grade
9
Congratulations, you passed.
```

6.3.2   The Instruction if-else-end

A second format for the instruction if is to add a statement else. This new statement allows programmers the use of a single if statement to test either one of two sets of conditions. The new if instruction is known as an if-else-end instruction. Its format is:

```
if condition
    set of instructions a1
else
    set of instructions b1
end
```

If the condition is valid then its value is **true** and the set of **instructions a1** is executed. After doing this, the program continues with the instructions following the **end** keyword. If the condition is not valid, that is, it is **false**, then the set of **instructions b1** is executed. When the last one of the instructions in set b1 is executed, the program continues with the first instruction after the **end** keyword. We can readily see that the statement if-else-end is equivalent to two if-end statements. In any case, the programmer has the freedom to use the instruction he/she thinks is more convenient.

Example 6.2 Use of the instruction if-else-end
We rewrite Example 6.2 using the statement if-else-end:

```
% File six_2.m
% Reads in a grade and writes out
% if the student passes or fails.
```

```
% Reads in the grade
calif = input ('Enter the grade:\n');
if calif >= 7.5
    fprintf ('Congratulations. You passed.\n')
else
    fprintf ('I am sorry. You failed.\n')
end
```

The program behaves in a similar way to Example 6.1.

6.3.3   The Instruction `elseif`

A third form of the statement `if` uses the keyword `elseif`. With it we can check for a greater number of conditions. We can have as many `elseif` instructions as needed. This is the case of having several conditions to check. There may be an `else` instruction if necessary. The format is:

```
if condition_1
    instructions A1
elseif condition_2
    instructions A2
elseif condition_3
    instructions A2
...
elseif condition_n
    instructions An
else
    instructions B
end
```

The instruction `if-elseif-end` checks which condition is first satisfied and executes the instructions that follow that condition. After that, it continues with the instructions following the end keyword.

Example 6.3 Use of the instruction `if-elseif-end`
We now modify Example 6.1 to include a statement if-else-end. This instruction can be used in the previous script to display an error if the user enters a letter instead of a numeric grade. We can correct that with the following program:

```
% File six_3.m
% % Reads in a grade and writes out
% if the student passes or fails.
%
% Asks for the grade
```

```
calif = input ('Enter the grade:');
if calif >= 7.5
    fprintf ('Congratulations. You passed.')
elseif calif < 7.5
    fprintf ('I am sorry. You failed.')
else
    fprintf ('Error-You must enter a numeric grade')
end
```

6.3.4 The Statement switch-case

The statement switch-case can be used when we wish to check if an expression is equal to some value and it cannot be used to check inequality conditions such as a > 5 or b < 3. However, it finds a great deal of applications where it is required to execute a set of instructions if a condition is equal to a given value. The format is:

```
switch expression
    case value 1
        Statements a1
    case value 2
        Statements b1
    otherwise
        Statements c1
end
```

The variable value can be a numeric quantity or a string. Furthermore, each case may have one or more values which can be enclosed within curly braces, like in:

```
case {value_1, value_2,..., value_k }
    statement 1
        ⋮
    statement m
```

In this case, if the expression is equal to any of the values value_1, value_2,..., value_k, the statements 1 to m are executed.

Example 6.4 Use of the instruction switch-case
As an example, we can modify the program from Example 6.1:

```
% File six_4.m
%
```

```
% Reads in data and writes out the result
% indicating if the student passes or fails.
%
% Enter the grade
calif = input ('Enter the grade:\n');
switch calif
   case {7.5, 8, 8.5, 9, 9.5, 10}
      fprintf ('Congratulations. You passed.\n')
   case {0, 0.5, 1, 1.5, 2, 2.5, 3, 3.5, 4, 4.5, 5}
      fprintf ('I am sorry. You failed.\n')
   otherwise
      fprintf ('It was not enough. You failed.\n')
end
```

6.3.5   The Instruction `for`

The instruction `for` is used to repeat sets of instructions that have to be repeated a number of times. They are also known as `for` loops because the set of instructions is repeated a number of times as in a loop. The format for this instruction is:

```
for x = values
   statement_1
   statement_2
      ...
   statement_n
end
```

If `x` attains the `values`, then `statements` 1 to `n` are executed. The next statement to be executed is the one following the `end` keyword. It is possible that `statements` 1 to `n` are never executed. This is the case when the last value in the list of values is smaller than the initial value.

Example 6.5 Use of the Instruction `for`
We wish to add the first 10 integer numbers. The following m-file does this task:

```
% File six_5.m
%
sum = 0;
for i = 1:10
   sum = sum + i;
end
fprintf ('The result is %g.', sum)
```

But if we only wish to add even integers between 1 and 10:

```
% File six_5a
sum = 0;
for i = 0: 2: 10
    sum = sum + i;
end
fprintf ('The result is %g \n', sum)
```

Note that 1:10 means that variable i varies from 1 to 10 in increments of 1 in each iteration. This can also be written as 1 : 1 : 10. For the second program, to select the even integers we use 0 : 2 : 10 which indicates that variable i changes value in increments of 2 starting with 0 and going up to 10. If required, the increment can be negative as in 4 : - 1 : 0 which means that it starts at 4 and it is decremented by 1 and ends at 0. Finally, the decrement can be a fraction, as in 4: 0.25 : 5. This would give 4, 4.25, 4.50, 4.75, and 5.

Example 6.6 Factorial function
The factorial of an integer n is defined as

$$n! = 1 * 2 * \cdots * n$$

We can use a **for** loop to find the factorial. This can be done with the following m-file:

```
% File six_6.m
% Factorial of a non negative integer.
%
n = input ('Enter a non-negative integer: ');
n_factorial = 1;
for i = 1: n
    n_factorial = n_factorial*i;
end
fprintf('The factorial of %g is %g.', n, n_factorial)
```

Because MATLAB has an internal function to evaluate the factorial of a non-negative integer named `factorial`, we have named our variable `n_factorial`. In general, users MUST NOT use names of MATLAB functions as variables or function names.

6.3.6   Nested Loops

We can nest **for** loops. That is, to embed a **for** loop inside another **for** loop. We may nest as many **for** loops as we wish.

Example 6.7 Use of nested `for` loops
We wish to add all the elements a_{ij} in an $n \times m$ matrix. This can be done with the double sum:

$$sum = \sum_{i=1}^{m}\sum_{j=1}^{n} a_{ij}$$

This equation can be implemented with the following file:

```
% File six_7.m
% This m-file evaluates the sum of
% the elements of a matrix n x m.
n = input ('Enter the number of rows\n');
m = input ('Enter the number of columns\n');
% Reads in the elements.
% Initialize the sum.
sum = 0;
% Reads in the matrix elements and adds them.
for i = 1: n
   % Reads in the elements of row i and adds them.
   for j = 1: m
      fprintf('Enter the matrix element %g,% g', i, j);
      a(i, j) = input('\n');
      sum = sum + a(i, j);
   end
end
fprintf('The total sum is %g\n', sum)
```

6.3.7   The `while` Loop

The `while` loop is used to repeat a set of instructions for an unknown number of times. The difference between a `while` loop and a `for` loop is that the `for` loop has to be repeated a known number of times. The format for the `while` loop is:

```
while condition
   statement_1
   ...
   statement_n
end
```

A `while` loop works as follows: If the `condition` does not hold, the instructions after the `end` keyword are executed. If the `condition` holds, then `statements_1` to n are executed. At this point, the `condition` is tested again. If it still holds, then the process is repeated again and `statements_1` to n are

executed. If at any time the condition is tested and it does not hold, then the
next executed instruction is the first one after the **end** keyword.

Example 6.8 Use of the while loop
We wish to evaluate the volume of spheres with radii 1 to 5 in increments of
1. We can do this with the following program:

```
% File six_8.m
%
% It evaluates the volume of spheres with radii 1 to 5.
%
% r is the radius of the sphere.
r = 0;
while r < 5
    r = r + 1; % The increment is equal to 1.
    vol = (4/3)*pi*r^3;
    fprintf('The radius is %g and the volume is %g.\n', r, vol)
end
```

When we run this file six_8 we obtain,

```
>> six_8
    The radius is 1 and the volume is 4.18879.
    The radius is 2 and the volume is 33.5103.
    The radius is 3 and the volume is 113.097.
    The radius is 4 and the volume is 268.083.
    The radius is 5 and the volume is 523.599.
```

But if we change the increment from 1 to 0.5, then we repeat the loop 10
times.

6.4 Functions

A function in MATLAB is a subprogram that can be used by a main larger
program or by another function. It is usually used to perform repetitive tasks.
The format of a function is:

```
function y = operation(x)
    statements of the function
    y = operation
end
```

Here, x is the argument or input parameter, y is the output parameter, and
operation is the name of the function. Note that y = operation is the last

instruction in the function definition. Additionally, we can add an **end** instruction, but this is optional if the function is saved in a separate file. A function can be called by another function. It can be called by itself in a recursive fashion. As an example, let us use a function to evaluate the factorial of an integer number. We include checks in the case that the number is not an integer or it is a negative integer.

Example 6.9 Factorial evaluation using a function

We wish to evaluate the factorial of an integer. In case that the number is not a positive integer, then it writes a message indicating this case. The function to evaluate the factorial is:

```
function x = factorial1 (n)
   factorial1 = 1;
   if n == 0
      factorial1 = 1;
   else
      for i = 1: n
         factorial1 = factorial1*i;
      end
   end
   x = factorial1;
end
```

If we execute this function for n = 3, 4 and 5 we obtain

```
>> factorial1(3)
     ans =
     6
>> factorial1(4)
     ans =
     24
>> factorial1(5)
     ans =
     120
```

Note that we did not check for n to be a non-negative integer.

Example 6.10 Use of a function inside another function

We now show another evaluation of the factorial. Here we wish to write an m-file that evaluates the factorial and checks if the number n is an integer and if it is negative. The file is

```
% File six_10.m
% Evaluation of the factorial of an integer.
```

```
% It includes messages in the case the number given
% is a non-negative integer.
n = input ('Enter an integer number n: \n');
error = 0;
if floor(n) ~= n % checks if n is an integer number.
   % floor(n) Evaluates the integer part of n.
      error = 1; % n is not an integer number.
end
if n < 0 % checks if n is a negative number.
   error = 2; % n is a negative number.
end
if error == 1
   fprintf('The number entered is not an integer one.\n')
elseif error == 2
   fprintf('The number entered is a negative one. \n')
elseif error ~= 1 & error ~= 2
   x = factorial1(n);
   fprintf('The factorial of %g is %g.\n', n, x )
end
```

If we run this file we obtain the following:

```
>> six_10
   Enter an integer number n
      -6
   The number entered is a negative one.
>> six_10
   Enter an integer number n
      4
      The factorial of 4 is 24.
```

Thus, the factorial of 4 is 24. We run again the m-file

```
>> six_10
   Enter an integer number n
   1.4
   The number entered is not an integer one.
```

Example 6.11 Solution of a second order equation
The solutions of the quadratic equation

$$ax^2 + bx + c = 0$$

are given by

$$x_{1,2} = \frac{-b \pm \sqrt{b^2 - 4ac}}{2a}$$

A function that calculates these solutions is quadratic (a, b, c) saved in the file quadratic.m:

```
function [x1, x2] = quadratic(a, b, c)
    % This function evaluates the roots x1, x2
    % of the quadratic equation ax∧2 + bx + c = 0
    % The file is quadratic.m
    discriminant = b∧2 - 4*a*c;
    x1 = (-b + sqrt(discriminant))/(2*a);
    x2 = (-b - sqrt(discriminant))/(2*a);
```

When we run this function we get the solutions for the equation

$$x^2 + 2x + 3 = 0$$

```
>> [x1, x2] = quadratic(1, 2, 3)
   x1 =
   -1.0000 + 1.4142i
   x2 =
   -1.0000 - 1.4142i
```

And for equation

$$4x^2 + 8x - 60 = 0$$

```
>> [x1, x2] = quadratic(4, 8, -60)
   x1 =
        3
   x2 =
       -5
```

6.5 Variables of Functions

The variables of the main program are independent of the variables of a function, even though they may have the same name. In general, they have different numeric values, unless they are declared as function arguments. To see this, let us consider the function Test in the following example.

Example 6.12 Variables in a function
In this example we check how variables are passed between subroutines and functions. Variables are not declared as input/output arguments.

```
function y = Test(x)
% Test function to watch variables.
fprintf ('\nJust entered Test. \n')
a = 1; b = 10; c = 100;
y = 3;
fprintf('I am inside the function Test. \n')
fprintf ('Variables are:\n a = %g, b = %g, c = %g', a, b, c)
fprintf ('\nExiting the function Test. \n')
```

If this function is called from the program six_12

```
% File six_12.m
a = 4;
b = 2;
c = 0.47;
fprintf('Variables before Test are:\n a = '
fprintf('%g, b =%g, c =%g', a, b, c)
p = Test(1);
fprintf ('I am back in the main program \n');
fprintf ('Variables are:\n a = %g, b = %g, c = %g\n', a, b, c)
```

When we run this program we obtain:

```
>> six_12
   Variables before Test are:
   a = 4, b = 2, c = 0.47
   Just entered Test.
   I am inside the function Test.
   Variables are
   a = 1, b = 10, c = 100
   Exiting the function Test.
   I am back in the main program
   Variables are
   a = 4, b = 2, c = 0.47
```

From this run we see that the values of a, b and c given before entering the function Test are not passed to the function. In addition, the variable values for a, b and c defined inside Test are not passed to the main program. Thus, in case we need to pass them either to the function or to the main program we need to declare them as function arguments as in the following example:

Example 6.13 Passing variables as function arguments

We now show how we can pass variables from the main program to functions and vice versa, as function arguments. Let us consider the following function where we have added as arguments the variables a, b and c.

```
function [a, b, c] = Test13(a, b, c)
  fprintf ('Variables in Test13 are:\n')
  fprintf ('a = %g b = %g c = %g', a, b, c)
  a = a*2;
  b = b*10;
  c = c*100;
  y = [a, b, c];
  fprintf('\nI am inside Test13.\n')
  fprintf('Variables have been changed to\n')
  fprintf('a = %g b = %g c = %g', a, b, c)
  fprintf('\nExiting the function.')
```

And we use it in the main program six_13:

```
% File six_13.m
a = 4;
b = 2;
c = 0.47;
fprintf('Variable values before Test13 are:\n')
fprintf('a = %g b = %g c = %g\n', a, b, c)
[a, b, c] = Test13(a, b, c);
fprintf('\nBack in the main program\n');
fprintf('Variables are a = %g b = %g c = %g', a, b, c)
```

When this program six_13.m is run, we get the following results:

```
>> six_13
   Variable values before Test13 are:
   a = 4 b = 2 c = 0.47
   Variables in Test13 are:
   a = 4 b = 2 c = 0.47
   I am inside Test13.

   Variables have been changed to
   a = 8, b = 20, c = 47
   Exiting the function.
   Back in the main program
   Variables are:
   a = 8 b = 20 c = 47
```

We note the following: At the beginning of the program the values for the constants are a = 4, b = 2, and c = 0.47. These values are passed to the function Test13 and they are displayed first after the text line Variables in Test13 are:. Then, they are changed and displayed from within the function Test13 as:

a = 8, b = 20, and c = 47

Then we exit the function Test13 and return to the main program where the variable values are displayed. We see that the values now are the same that were changed in the function Test13.

This way to pass variables is called **call by value**. We may pass them to a function but let them unchanged in the main program.

6.5.1 Global Variables

In the previous section we saw how we can pass variables between a main program and a function. We learned that values do not pass just by giving the same name to variables. They need to be passed by value in order to be used in another function or in a main program. Variables defined in this way are called local variables. This may be convenient sometimes when we wish to use the same name for different variables, and we do not wish to change the variable values. When we define a variable, MATLAB assigns it as a local variable. Unless the programmer wishes to use in another function with the same value, we have to remove the local variable restriction and declare it as a global variable. In this way, every time we refer to a given variable previously declared as a global one, it is going to have the same value, and if this value is changed, it is going to change in any other function and main program where the variable is declared as global. This means that if a function does not declare it as a global variable, it will not have the global variable value in that function. The instruction to declare a variable as global is

```
global a b
```

This means that variables a and b are global variables.

Example 6.14 Program with a global variable
To see how global variables work, let us consider the variable a = 3. We wish to have this variable in a function and in a main program. The following file six_14.m, which includes the script and the functions change1 and change2, shows this:

```
% File six_14.m
% Program to check local and global variables.
```

```
global a
a = 3;
fprintf('Value of ''a'' before function change1 a = %g ', a)
change1
fprintf('Value of ''a'' after function change1 a = %g \n', a)
change2()
fprintf('Value of ''a'' after exiting change2 a = %g \n', a)

function x = change1()
  global a
  fprintf('Value of ''a'' after entering change1 a = %g.\n', a)
  a = 7;
  fprintf('Value of ''a'' modified in change1 a = %g.\n', a)
end

function x = change2()
  fprintf('Value of ''a'' after entering change2 a = %g.\n', a)
  a = 12;
  fprintf('Value of ''a'' modified in change2 = %g.\n', a)
end
```

Now we run the file six_14,

```
>> six_14
  Value of ''a'' before entering function change1 a = 3.
  Value of ''a'' before after entering function change1 a = 3.
  Value of ''a'' modified in change1 a = 7.
  Value of ''a'' after returning from function change1 a = 7.
  Value of ''a'' entering change2 a = 7.
  Value of ''a'' before exiting change2 a = 12.
  Value of ''a'' after returning from change2 a = 7.
```

As we see in the main program and in the functions, the variable a is global in the main program and in function change1 and it is only local in change2. When we run the program, the value of a is passed to a in the instruction global a. There it keeps the same value until we change it to a = 7. It passes this value to the main program using again the instruction global a. The value of a is passed as an argument in the function change2. There we change its value to a = 12. When we exit this function, the value of a is not changed in the main program because it is a local variable in change2.

6.5.2   The Instruction return

In general, a function ends with the last instruction, but sometimes we need to end the function before the last instruction and **return** to the main program or function that called the function. We can do this with the instruction **return**. When the sequence of instructions finds a **return**, it ends the function and it exits it, returning to the function or program that called it. We present an example to show how the instruction **return** works.

Example 6.15 Use of return in a function

Let us consider the script `six_15.m` which uses the function `greater_smaller` to find out if a number is greater than or smaller than 0.

```
% File six_15.m
x = input ('Enter the value of x: \n');
greater_smaller (x);
fprintf (' The run ends. \n')

function greater_smaller (x)
  if x > 0
    fprintf (' x is less than 0. \n')
  elseif x > 0
    fprintf (' x is greater than 0. \n')
    return
  else
    fprintf (' x is equal to 0. \n')
end
```

Now we have several runs to show how the instruction **return** works.

```
>> six_15
  Enter the value for x:
  0
  x is equal to 0.
  The run ends.
>> six_15
  Enter the value of x:
  3
  x is greater than 0.
  The run ends.
>> six_15
  Enter the value of x:
  -3
  x is less than 0.
  The run ends.
```

We observe the following: When x is greater than or smaller than 0, the function finds a **return** instruction and interrupts it before executing the remaining instructions. The control is passed to the main program which prints **The run ends.** If x = 0 the function ends the **if** statement and then goes to the main program. In this last case the function is executed to the last instruction.

6.5.3   The Instructions nargin and nargout

The instructions **nargin** and **nargout** are used to find out the number of input and output arguments in a function, respectively. For example, for the function in Example 6.15, we have **nargin** = 1, **nargout** = 1. **nargin** is the acronym for Number of ARGuments in the INput and **nargout** for Number of ARGuments in the OUTput. **nargin** and **nargout** are variables.

These two instructions are useful when we have a function that allows the branching to different parts of it depending on the variables when calling a function. For example, a function to solve a quadratic equation, but where the user only gives two coefficients might branch to a section of the function to solve a first order equation.

6.5.4   Recursive Functions

A recursive function is a function that calls itself. This property is available for MATLAB functions. We show an example to show the recursivity in MATLAB functions.

Example 6.16 Recursive evaluation of the factorial function
The simplest example of a recursive function is the factorial function. As we already know, the factorial of a non-negative integer n is defined as

$$n! = 1 * 2 * ... * n$$

This can be rewritten as
$$n! = n \times (n-1)!$$

In Example 6.15 we saw that the factorial is evaluated by the function **factorial1**. Using recursion the function can be written as:

```
function x = fact_rec(n)
if n >= 1
   x = n*fact_rec(n-1);
else
   x = 1;
end
```

When we run this program we get

```
>> fact_rec(6)
   ans =
   720
```

We see that the function is calling itself and producing the expected result.

6.6 File Management

Up to this point, input data has been entered through the keyboard. The keyboard is thus an input device. The results are usually displayed on the `Command Window` or in a `Figure` window. Thus, the computer screen is the output device. Another way to give input data is by using a file where we have somehow stored the input data. This file can be created by MATLAB or by any other computer program. This last option allows data exchange between MATLAB and any other software package. For example, a spreadsheet such as Excel might exchange data with MATLAB, so they can be processed and visualized. In the same way, data generated in MATLAB can be used by other programs. An obvious advantage of saving data in a file is that we can use it in a later MATLAB session or we can use it in another computer or send it to another user.

TABLE 6.3: Permit codes to open files

Permission	Action
'r'	Opens the file to read. The file must already exist. If it does not exist, it sends an error message.
'r+'	Opens the file to read and write. The file must already exist. If it does not exist, it sends an error message.
'w'	It opens the file to write. If the file already exists, it deletes its contents. If it does not exist, it creates it.
'w+'	It opens the file to read and write. If the file already exists, it deletes its contents. If it does not exist, it creates it.
'a'	It opens the file to write. If the file already exists, it appends the new contents. If it does not exist, it creates it.
'a+'	It opens the file to read and write.

6.6.1 File Opening and Closing

To read or write data to a file we have first to open it. MATLAB can open a file using the instruction `fopen` that has the format:

```
fid = fopen (file name, permissions)
```

TABLE 6.4:   File handles

Handle fid	Meaning
-1	Error when opening the file. Usually, when we want to read from a non-existent file.
0	Standard input. Usually, the keyboard is always open to write ('r').
1	Standard output. Usually, the Command Window. Always open with permission to add data at the end of the existing one.
2	Standard error. Always open with permission to add data at the end of the existing one.

Here, `file name` is the file name and it must exist in order to open it. If we want to write to a non-existing file, the file is created and then the data is written in. `Permissions` is a variable that specifies how the data is written to the file. The permission codes available are given in Tables 6.3 and 6.4. We can open as many files as we wish. `fid` is the handle that is used to identify the file. The handle values start with 3. Handle values from -1 to 2 have a special use in MATLAB and they are given in Table 6.4. After using the data from a file, we must close it. The instruction to close files is `fclose`. The format is:

```
status = fclose(fid)
```

The variable `fid` has the handle's value for the file we wish to close. The variable `status` is another handle that indicates if the file was successfully closed. The result `status = 0` indicates that it was closed, while `status = -1` indicates that the file was not closed.

Example 6.17 Use of `fopen`.
Let us suppose that we want to open a file called `Example_six_17.txt`. Then we use:

```
handle_file = fopen('Example_six_17.txt', 'r');
handle_file
handle_file =
-1
```

The value of `handle_file` is -1 because the file does not exist. If we use 'w' instead of 'r', we get:

```
handle_file = fopen('Example_six_17.txt', 'w');
handle_file
handle_file =
3
```

The new handle value is 3 indicating that before writing to file Example_six_17.txt, the file had to be created because it did not exist before. We now create another file data.txt with:

```
handle_file2 = fopen('data.txt', 'w')
handle_file2 =
4
```

The handle for this file is 4 because handle values are sequentially assigned. In the Current Folder window we see that these files were created. We can use any word processor, such as Office Word, the Notepad or the WordPad, and even we can open a text file with the MATLAB editor, to see their contents.

To create a file in another directory different from the one we are working on we only have to specify the path. For example,

```
File_name = fopen('C:\new\file.txt', 'w')
File_name = 5
```

To close a file we use the instruction fclose whose format is

```
status = fclose(fid)
```

fid is the handle corresponding to the file we wish to close. The value of the variable status indicates if the file was closed successfully (status = 0). If this was not the case and the file was not closed, MATLAB sends an error message to let us know that the file could not be closed. Before closing MATLAB, it is convenient to close all files open during the session. This avoids losing the data stored in them during the session.

6.7   Writing Information to a File

The simplest way to write information to a file is with the instruction fprintf. We have used this instruction before to display output data in the computer screen through the MATLAB Command Window as:

```
fprintf ('Display this text to the screen')
Display this text to the screen
```

To write to a file we can use this same instruction. Let us create two new files with,

```
handle1 = fopen('file1.txt', 'w');
handle2 = fopen('file2.txt', 'w');
```

To write to a file we need to use its handle as in:

```
fprintf (handle1, 'We write here to file1.txt \n');
fprintf (handle2, 'We now write here to file2.txt \n');
```

To see what is written in them we can open them either with the Notepad or Worpad or any text editor, but first we close them with:

```
fclose(handle1);
fclose(handle2);
```

6.7.1   Reading and Writing Formatted Data

MATLAB allows users to read and write formatted data to a file. We show the procedure with an example.

Example 6.18 Writing formatted data

Suppose we want to write data about the planets in the solar system. The data we wish to write is:

1. Name 7-character string.

2. Position 2-digit integer number.

3. No. of moons 2-digit integer number.

4. Diameter 10-digit floating number.

If data in a planet name is less than the seven characters indicated, the remaining characters are filled with blank spaces. The elements of each item are stored in an array. The data is:

```
Planet name = ['Mercury'; 'Venus  '; 'Earth  '; 'Mars   '; ...
    'Jupiter'; 'Saturn '; 'Uranus '; 'Neptune']
Position = [ 1; 2; 3; 4; 5; 6; 7; 8];
No_of_moons = [ 0; 0; 1; 2; 63; 34; 21; 13];
Diameter_in_km = [ 4880; 12103.6; 12756.3;...
    6794; 142984; 120536; 51118; 49532];
```

Note that we are writing the data as column vectors. To see the data for the second planet we write:

```
>> fprintf ('%s\n%g\n%g\n%g\n', Name(2,:), Position(2,:),...
     No_of_moons(2,:), Diameter_in_km(2,:));
```

To obtain:

```
Venus
2
0
12103.6
```

To write this information to a file we use the following script:

```
% File Example_6_18.m
%
handle_planets = fopen('Planets.txt', 'w');
for i = 1 : length(Position);
   fprintf(handle_planets,'%7s ,%5d ,%2d ,%2d \n', ...
     Planet_name(i, :), Position(i, :), ...
     No_of_moons(i, :), Diameter_in_km(i, :));
end
fclose(handle_planets);
```

We can now read the data in a file with a script or a function written in m-language. We have to check for several things when reading this data:

1. If we have gotten to an **end_of_file** which has the variable name **feof**.

2. Read each string with **fscanf** and assign it to its corresponding field.

3. Close the file after we find the **feof**.

The m-file is:

```
% This is file read_data.m
%
handle_data = fopen('Planets.txt', 'r');
% We define the names of the column vectors.
Names = [ ];
Positions = [ ];
Moons = [ ];
Diameters = [ ];
while ~feof(handle_data)
   % Read the planet name
   stringg = fscanf(handle_data, '%7c', 1);
   Names = [Names; stringg];
   comma = fscanf(handle_data, '%1c', 1);
```

```
    % Read the position
    number = fscanf(handle_data, '%5d', 1);
    Positions = [Positions; number];
    comma = fscanf(handle_data, '%1c', 1);
    % Read the number of moons
    number = fscanf(handle_data, '%5d', 1);
    Moons = [ Moons; number];
    comma = fscanf(handle_data, '%1c', 1);
    % Read the diameter
    number = fscanf(handle_data, '%12e', 1);
    Diameters = [ Diameters; number];
    end_of_line = fscanf(handle_data, '%1c', 1);
end
fclose(handle_data);
```

The instruction **while** checks for the **end_of_file**. The instruction **fscanf** searches for the planet name characters using the format %7c. It reads the first seven characters including the blank spaces. If we have used instead the format %7s, only the characters are read and the blank spaces are ignored. The 1 after %7c means that only an element is read. In this way, **stringg = fscanf(handle_data, '%7c', 1)** does the following: reads an element of the open file which has the handle **handle_data** and places it in the variable **stringg**. The line **comma = fscanf(handle_data, '%1c', 1)** indicates that after reading the first variable, a comma is read. We do not do anything with the comma but we need to read it. Otherwise it is read by the next instruction **fscanf**.

For the variables **Positions**, **Moons**, and **Diameters**, we need to read a numerical value. For this we use **number = fscanf(handle_data, '%5d', 1)**. After reading the data, we arrive at the end of the line. This is read with a special character **end_of_line = fscanf(handle_data, '%1c', 1)**. Again, we do not need this character but we need to read it, for the same reason we did with the comma. To see how this file works, we run the file and see the variables:

```
>> read data
>> who
```

We now see the data:

```
>> Names
Names =
Mercury
Venus
Earth
Mars
```

```
Jupiter
Saturn
Uranus
Neptune

>> Positions
Positions =
1
2
3
4
5
6
7
8

>> Moons
Moons =
0
0
1
2
63
34
21
13

>> Diameters
Diameters =
1.0 e+005 *
0.0488
0.1210
0.1276
0.0679
1.4298
1.2054
0.5112
0.4953
```

As we can see, these are the values entered before.

TABLE 6.5:   Options for variable precision

Option	Meaning
'char'	8-bit characters. Used for text.
'short'	16-bit integers. Integer numbers in the range from -215 to 215-1.
'long'	32-bit integer (2's complement). Integers in the range from -231 to 231-1.
'ushort'	Unsigned integer (16 bits).
'uint'	Unsigned integer (32 bits).
'float'	Single precision floating point real numbers (32 bits).
'double'	Double precision floating point real numbers (64 bits).

6.7.2   Reading and Writing Binary Files

So far, we have used alphanumeric data that can be read by any text processor besides the MATLAB editor. This type of data is called ASCII data. A disadvantage of this type of data is the size of the files when we handle a large amount of data. An alternative way is to store data in a binary format. This is a more efficient way to store information. Unfortunately, information stored in a binary format cannot be read by a text processor. To write and read in a binary format we use the instructions `fwrite` and `fread`, respectively. The format for `fwrite` is:

```
count = fwrite(handle, A, 'precision')
```

The variable A contains the data to be written. `count` is the number of elements that were successfully written. `handle` is the handle for the opened file. `precision` gives information about how we want to store the information. Some of the options for this variable are given in Table 6.5.

Example 6.19 Reading and writing binary data
Let us suppose that we want to write the following data in binary format:

$$A = \begin{bmatrix} 57 & 10 \\ 14 & 75 \end{bmatrix}$$

$$B = 27, \ C = \text{'MATLAB'}$$

This data is entered as:

```
>> A = [57 10; 14 75];
>> B = 27;
```

```
>> C = 'MATLAB';
```

Then we open the file

```
>> fid_binary = fopen('binary.dat', 'w')
   fid_binary = 3
```

To write we use

```
>> fwrite(fid_binary, A, 'double')
   ans =
   4
```

The answer is 4 indicating that 4 elements were written. These elements belong to matrix A. We now write in B:

```
>> fwrite(fid_binary, B, 'short')
   ans =
   1
```

This time the result is a 1 because B has only one element. We now continue with C:

```
>> fwrite(fid_binary, C, 'char')
   ans =
   6
```

The answer is 6 because the word MATLAB has 6 elements. We now close the file

```
>> fclose(fid_binary)
```

Now we try to read the file with the WordPad and we get the results shown in Figure 6.2. As we can see, the data is not what we wrote to the file because it is in binary format and not in ASCII format, thus trying to read it with a text processor does not display the information stored in it. To read the data, first we open the file:

```
>> fid_bin = fopen('binary.dat', 'r')
   fid_bin = 3
```

And now we use **fread**, whose format is:

```
>> [A, count] = fread(fid_binary, [2, 2], 'double')
```

FIGURE 6.2: WordPad window showing the contents of `binary.dat`.

where A is the matrix name and `count` is the number of elements (4 in the case of matrix A). If A were an 8 × 4 matrix, then instead of [2, 2] we should write the size [8, 4]. The results are:

```
>> [A, count] = fread(fid_bin, [2, 2], 'double')
>> A =

57 10
14 75
count =
4
```

Data must be read in the same order and with the same format that it was written. To read out the remaining variables we use:

```
>>  B = fread(fid_bin, [1], 'short')
   B =
   27

>> C = fread(fid_bin, [6], 'char')
   C =
   77
   65
   84
   76
   65
   66
```

Note that C is an ASCII string in a column vector. To change it to a row vector in characters we transpose it and then use `setstr` as in:

```
>> C = setstr(C')
   C =
   MATLAB
```

Finally, we close the file:

```
>> fclose(fid_bin)
   ans = 0
```

This value indicates that the file was successfully closed.

6.8 Passing Data Between MATLAB and Excel

A software package used in engineering, science, and finance is Excel. Excel and MATLAB can read and write data to files. In this section we show how such files can be used by any of the packages. For example, MATLAB can write data separated by commas in files with extension csv, for comma separated values, and then read by Excel, and vice versa.

6.8.1 Exporting Data to Excel

To show how we can export data from MATLAB to Excel we have the following example.

Example 6.20 Exporting data to Excel from MATLAB
Let us consider the following data about the five countries with the largest territories in square miles in the American continent together with their capital cities:

```
Canada, Ottawa, 3849660
United States of America, Washington D.C., 3787319
Brazil, Brasilia, 3300410
Argentina, Buenos Aires, 1073596
Mexico, Mexico D.F., 759589
```

This data is written by MATLAB to file `countries.csv`. The following file opens the file `countries.csv`, writes the data, and closes the file. Each country name must have the same number of characters. Each capital name must have ten characters, including blank spaces.

FIGURE 6.3: Data in the file countries.csv.

```
% File Example_6_20.m
Country = ['Can'; 'USA'; 'Bra'; 'Arg'; 'Mex']
Capital=['Ottawa ';'Washington';'Brasilia ';'B. Aires ';'MexicoCity']
Size = [3849660; 3787319; 3300410; 1073596; 759589]
handle = fopen('countries.csv', 'w')
for i = 1: 5
    fprintf(handle, '%10s, %10s, %7d \n',...
        Country(i, :), Capital(i, :), Size(i, :))
end
fclose(handle)
```

Now we open the file `countries.csv` with the Wordpad and we see the contents shown in Figure 6.3. We note that it is in comma-separated-values format. The commas are called separators or delimiters. Now we proceed to open the file with Excel. Since the file was not created by Excel, it has to be imported to Excel. The `Import Wizard` is automatically opened. It consists of three windows. The first window is shown in Figure 6.4. Here we indicate that the data is delimited by commas as indicated. After pressing the `Next` button the second window for the `Import Wizard` opens and here we indicate that the data is delimited by commas as shown in Figure 6.5. Finally, when we press the `Next` button we get to the third window in the `Import Wizard`. Here we select the columns we want to import and set the data format, as shown in Figure 6.6. Finally, the data is shown in Excel as can be seen in Figure 6.7.

FIGURE 6.4: Part 1 of the `Import Wizard`. Here we indicate that the data is separated by commas or tabs.

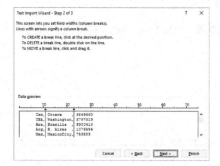

FIGURE 6.5: Part 2 of the `Import Wizard`. Here we indicate that the data is separated by commas.

FIGURE 6.6: Part 3 of the `Import Wizard`. Here we select each column and set the data format.

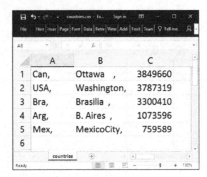

FIGURE 6.7: Data from `countries.csv` in Excel.

Another instruction used in MATLAB to write only numerical data to a file and read it by Excel is the instruction `csvwrite('file_name', m)`. For example, for the matrix A given by

```
>> A = [1957 10 5; 1950 10 8;1989 5 10 ]
```

```
A =
   1957 10 5
   1950 10 8
   1989 5 10
```

We can write this matrix to a file `list.csv` with

```
>> csvwrite ('list.csv', A )
```

The file `list.csv` can be readily opened with Excel.

6.8.2   Exporting Excel Files to MATLAB

The instruction `csv` also allows numerical data transfer from Excel to MATLAB. To show how this can be done, in Excel define the matrix A given by

$$A = \begin{bmatrix} 1 & 0 \\ 100 & 1 \\ 2 & 4 \end{bmatrix}$$

The matrix in Excel is shown in Figure 6.8. We save it in a file `Numbers.csv`.

Now, from the MATLAB `Command Window` we use the instruction `csvread` as `csvread('Numbers.csv')` to obtain the data in MATLAB as follows:

```
>> csvread('Numbers.csv')
```

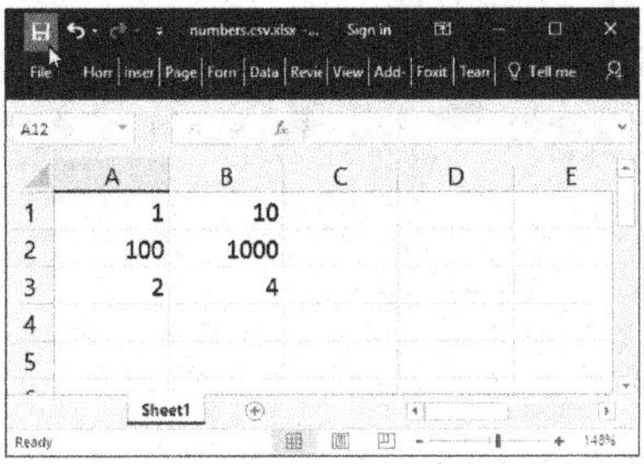

FIGURE 6.8: Data in Excel for `Numbers.csv`.

```
ans =
   1 10
 100 1000
   2 4
```

6.8.3 Reading Data from Excel Files

MATLAB can also read data from files with either extension `xls` or `xlsx`. To describe the procedure we use the Excel data shown in Figure 6.9. This data is saved in the file `years.xlsx` in the current directory for the MATLAB session. Now, in MATLAB we look at the `Current Directory` window and locate the file `years.xlsx`. We just double click on this file and then the `Import Wizard` opens requesting the variables to be imported. Selecting the variable to be imported (see Figure 6.10), it is displayed in the right-hand window and then we click on the `Finish` button to end the importing. In the `Workspace` window appears the variable which we can now use as any other variable created in MATLAB. This is shown in Figure 6.11.

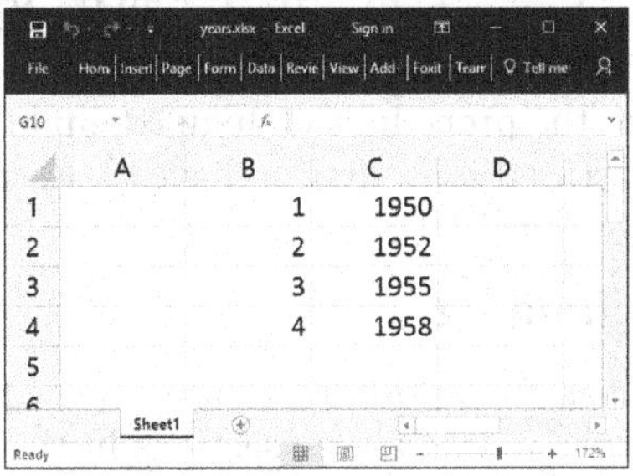

FIGURE 6.9: Data in Excel for `years.xlsx`.

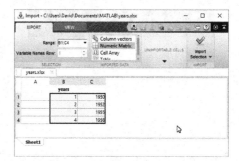

FIGURE 6.10: Import wizard for Excel files.

FIGURE 6.11: Import wizard for variables.

6.9   Publishing m-files from MATLAB

A very important part in programming is documentation. Sometimes this may be either a very tedious part in the programming process or a very easy one depending upon the programmer's style. Fortunately, in the case of m-language programming, MATLAB has a tool to create documentation once a program has been finished. This is known as publishing and the end result is a document that can be created in Word, HTML, XML, LaTeX, or Power Point. Furthermore, it is possible to run the program documented from the published file. We show the procedure with an example. In order to learn how an m-file has to be structured to it, first we have to describe cell programming.

6.9.1   Cell Programming

The MATLAB editor has the option to create cells. These cells are useful to run portions of the m-file and to create sections in the publishing process. A cell is a portion of an m-file having certain characteristics. A cell is composed of the following parts:

1. A beginning row which starts with a double percent sign followed by a space and a title text.

2. Comment lines that start with a percent sign followed by a space and text which is the body of the documentation.

3. Equations written in LaTeX style, and finally:

4. MATLAB instructions. We show the procedure with an example.

Example 6.20 Plotting of a sine function with cells

Let us suppose that we want to plot the sine function from 0 to 2π and then we want to modify the plot. To plot the function we use

```
x = 0: 0.01: 2*pi;
y = sin(x);
plot(x, y)
```

Once the function is plotted we add title, legends to the axes and a label with:

```
xlabel('x-axis')
ylabel('sine wave')
title('Plot of sin x')
```

Now we create the m-file using cells. In the first cell we place the first part of the m-file and in the second cell the m-file where we add the text part. We add comments to the m-file to make it self-explanatory. In the comment lines we leave a blank space between the percent sign % and the beginning of the text. The following m-file is in cell form:

```
%% Example of m-file using cells
%% Plot of a sine wave
% Here we plot a sine wave going from 0 to 2π
% As we know we have to create a vector for x values
% and then a vector for y values. The vector y has the
% information for the sine wave values
x = 0: 0.01: 2*pi;
y = sin(x);
plot(x, y)
%% Adding text information to a plot
% We add a title with title.
% We add a text to the x axis with xlabel.
% Finally, we add a text to y axis with yaxis.
%
xlabel('x-axis')
ylabel('sine wave')
title('Plot of sin x')
```

FIGURE 6.12: m-file divided in cells.

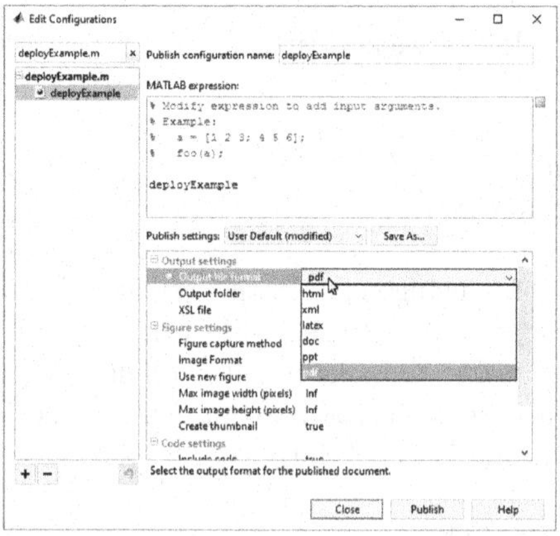

FIGURE 6.13: Publishing preferences.

The editor window is shown in Figure 6.12. We see that each cell is separated by a line and that each cell has different background color. To run the file in this mode, from the main menu we select the icon for **Run** and **Advance** and this runs the m-file a cell at a time. After running each cell we see the result of the second cell is the sine function plot, and the last cell result is the same plot with a title and with axes labels. The first cell does not display anything because there are no instructions in that cell. We now proceed to publish this m-file.

6.9.2   Publishing m-files

Now that we have the m-file in cell mode we can proceed to publish it. The result is a document in the format selected. The first step is to choose the **Publish** tab in the MATLAB m-file editor. In the **Publish** icon we can choose publishing in Word, HTML, XML, LaTeX, or Power Point. The default option is the HTML format. If we wish to change to any of the other formats we can do so in the `Publish` icon which displays the preferences window for publishing as shown in Figure 6.13. For the example we choose the default pdf format. In this window we can choose the button `Publish` . We can also do it with the icon `Publish` after we close the window for the preferences. This will start the publishing process. After a few seconds we get the pdf document shown in Figure 6.14. We see in the pdf document that it has

1. A title,

2. A table of contents,

3. MATLAB instructions,

4. Text explaining the instructions, and

5. Plots.

We now describe each part of the document:

1. The title "`Example of m-file using cells`" is the first line of the first cell (the first cell has only the title of the document).

   `%% Example of m-file using cells`

2. The table of contents is formed by the first line of each cell. That is, the lines with a double percent sign,

   `%% Adding text information to a plot`

   `%% Plot of a sine wave`

3. The MATLAB instructions are, for the second cell,

   ```
   x = 0: 0.01: 2*pi;
   y = sin(x);
   plot(x, y)
   ```

   And for the third cell

   ```
   xlabel('x-axis')
   ylabel('sine wave')
   title('Plot of sin x')
   ```

4. The text for each cell is the commented lines. For the second cell:

   `% As we know we have to create a vector for x values`

   `% and then a vector for y values.`

TABLE 6.6: Commands for MATLAB LaTeX

Traditional Equation	MATLAB LaTeX $$equation$$
a/b	\frac{ a }{ b }
a^2	a \wedge 2
$a_{k,n}$	a_{ k, n }
α^2	\alpha \wedge 2
$\sqrt{a+b}$	\sqrt{ a + b }
$\int (a+b)dt$	\int { (a + b)dt }
$a \leq b$	a \leq b
$a \geq b$	a\geq b
$\backslash$	\textbackslash

```
% The vector y has the information
% for the sine wave values.
```

For the third cell:

```
% We add a title with the instruction title.
% We add a text to the x axis with xlabel.
% We add a text to y axis with yaxis.
```

5. Finally, the plots are also displayed in the pdf document.

In the published document we may also include equations. They have to be written in the LaTeX format. Table 6.6 lists some of the more used MATLAB LaTeX formats to write equations. For example, to write

$$\int \sqrt{\alpha \sin(t)}\ dt$$

We use:

```
% $$ \int \sqrt {\alpha sin(t)}\, dt $$
```

Some rules have to be followed. These rules are:

1. If a row has an equation, there must be an empty comment row above and below.

2. There must be at least a blank space between the percent sign and the double dollar sign.

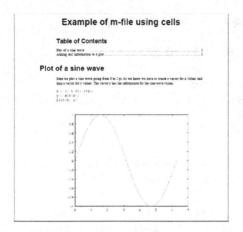

FIGURE 6.14: Top part of the deployed pdf document.

3. After a row with executable MATLAB instructions, the row has to start with a double percent sign, that is, a new cell has to start.

We now show an example to solve a quadratic equation.

Example 6.22 Publishing an m-file with equations
To solve the quadratic equation

$$ax^2 + bx + c = 0$$

We have the solutions:

$$x_{1,2} = \frac{-b \pm \sqrt{b^2 - 4ac}}{2a}$$

To publish these equations we have to write the m-file as

```
%% Solution of a second order equation
%% Introduction
% A second order equation of the form
%
% $$ ax^2  + bx + c = 0 $$
%
% has the solutions
%
% $$ x_{1} = \frac { -b - \sqrt{b^2 - 4ac}}{2a}$$
%
% $$ x_{2} = \frac { -b + \sqrt{b^2 - 4ac}}{2a}$$
%
%% Example
```

```
% As an example we solve the equation
%
% $$ 3x∧2 + 6x - 9 = 0 $$
%
% The data are then
a = 3; b = 6; c = -9;
%% Solutions
% The solutions are
%
x1 = (-b + sqrt(b∧2 - 4*a*c))/(2*a);
x2 = (-b - sqrt(b∧2 - 4*a*c))/(2*a);
%% Results
% Finally, we display the values of the roots. %
fprintf ('x1 is a root of the second order equation %g\n', x1)
fprintf ('x2 is a root of the second order equation %g\n', x2)
```

We publish to HTML and we get the document shown in Figure 6.15.

FIGURE 6.15: Top part of the published file in HTML format.

6.10   Concluding Remarks

MATLAB has integrated a powerful programming language called m-language. This language allows users to produce complex programs in a very short time when we compare it with other programming languages such as C, C++, Visual Basic, FORTRAN, among others. In the chapter we treated in detail several of the instructions needed to write a program in the m-language. The process was carried out through examples going from very simple to more complex examples containing programs. A treatment in input/output instructions was given, in particular, the case of reading to/from a file. Also the case of transferring information between MATLAB and Excel was seen. More advanced topics such as deployment of m-files to users not having a MATLAB license was also discussed and examples provided a good understanding of the topic. Finally, since program documenting, also known as publishing, is an important part of programming, MATLAB does also provide a tool for documenting m-files. The process can be used for publishing to other different formats, but only the process for an HTML document was used.

Chapter 7

Object-Oriented Programming

In this chapter a brief description is given on how to write programs using an object-oriented paradigm with MATLAB. Demonstrative examples are also presented.

7.1   Introduction

This chapter is a brief introduction to Object-Oriented Programming (OOP) as it is implemented in MATLAB. Although this way of programming is more associated with languages such as C++, Java, Python, among others, now it has been incorporated to MATLAB. A great deal of textbooks have been dedicated to the teaching of object-oriented programming, thus this section only presents the characteristics of OOP in MATLAB and gives the specifics to start programming using this paradigm. The chapter begins by defining the elements of OOP and continues with examples of how to program in MATLAB emphasizing the OOP.

7.2   The Object-Oriented Programming Paradigm

Object-oriented programming became popular in the 80s and has had a boom since 1990. It is used in high-level languages such as C++ and Java. One advantage of using OOP is that objects can be used to represent objects in a very clear way in real life. In this way OOP can be used to solve problems more clearly and easily, allowing the design of more efficient algorithms.

Object-oriented programming requires a clear and precise understanding of the following terms used in OOP and defined below:

- object

- attribute

- method

- class and instance

- encapsulation

- inheritance

- polymorphism

An **object** is a key element in OOP. It can represent a real object such as a bank account, a client or a transaction. Each object has a state and a behavior. The state of an object is the set of characteristics that define the object at a given instant of time. The behavior is described by the activities associated with the object.

Attributes or **properties** are the values that are stored internally by the object and that can be primitive data or other objects. The attributes or properties represent the state of the object.

A **method** is a set of programming instructions which are given a name. When a method is invoked, the program instructions are executed. The methods of an object define its behavior, for example, the actions that can be performed on it.

A **class** is the model or template from which an object is created. In this way the important features of the object are defined in both its state (or structure) and its behavior (set of methods). From a class, several different objects with similar characteristics can be defined. A specific example of a class is called an **instance** which in turn is called an object. Instances of the same class have the same type of attributes or characteristics and share the methods of the class.

To create **instances** of a class, it is necessary to include a constructor method in the class definition, which will receive a variable number of arguments to specify the initial conditions of the properties of the instance and return the constructed object with designated values.

To **encapsulate** means that information can be stored and that also can be worked with it so that status changes can only be made by the methods of the object and that no other object can change them. Encapsulation allows the hiding of the internal details of the implementation of these methods.

Some classes relate to each other through **inheritance**, forming a classification of hierarchy. Inheritance allows the definition and creation of new objects such as specialized types of preexisting objects. Objects inherit the properties and behavior of all classes to which they belong. New objects can share (and extend) their characteristics (attributes) and behavior (methods) without having to re-implement them.

Polymorphism is the property that allows OOP values of different types of data to be handled using a uniform interface. The same method can be used by several classes, but differently. When the same method is applied to different objects, although they share the same name of the method, the behavior will be produced for the type of object that is invoking it. When this occurs in "runtime", this feature is called late assignment (late binding) or dynamic allocation.

7.3 Classes in MATLAB

To create a class, in the MATLAB editor select from the Home menu New $\rightarrow$ Class as shown in Figure 7.1.

This opens a file with basic instructions on defining a class, as shown in the following list:

```
classdef Untitled
    % UNTITLED Summary of this class goes here
    % Detailed explanation goes here

    properties
    end

    methods
    end

end
```

As it can be seen, following the object-oriented paradigm in MATLAB, data and actions are combined into a single object called class. In the class definition (`classdef`), data is called properties and actions are called methods. A MATLAB class is defined in an m-file. This file should include a constructor, which is responsible for creating new objects and must have the same class name.

FIGURE 7.1: Creating a class in MATLAB.

In the definition of the properties of classes, MATLAB handles primitive data that can be of different type such as numeric, symbolic or string type and may also include instances of other classes. In MATLAB, the contents of them are handled as structures (**struct**), which were presented in Chapter 4.

The creation of a class and the main characteristics of OOP in MATLAB are shown through an example in which a class which handles fractions is created. It is also shown the definition of a class, including its constructor, the use of **setters** and **getters**, the mechanisms of inheritance and direct and indirect methods.

7.3.1 Creation and Use of a Class

Let us suppose that it is needed to calculate a fraction which consists of a numerator and a denominator. To avoid a division by zero, it will be assumed that the divisor will always be different from zero. Then a new class called My_Fraction is defined, for which the numerator and denominator are declared as their properties. The operation of calculating the fraction as the ratio of the numerator and denominator is declared as a first method called My_Fraction1. In MATLAB, methods are defined by functions (**function**) as shown below:

```
classdef My_Fraction1 < handle
   % This class obtains the ratio of two numbers.
   % It belongs to superclass handle.

   properties
      numerator = [ ]
      denominator = [ ]
   end
   methods
```

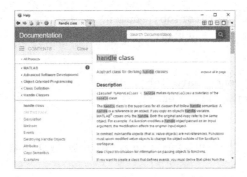

FIGURE 7.2: Help for viewing information about the superclass `handle`.

```
function Result = Fraction(obj)
    Result = obj.numerator/obj.denominator
  end
 end
end
```

If this class is now used to obtain the fraction, we have to create the object invoking an instance of the class `My_Fraction1` as follows:

```
>> a = My_Fraction1( );
>> a.numerator = 10; % numerator value.
>> a.denominator = 2; % numerator value.
>> a.Fraction
  ans =
    5
```

Before continuing, note that the class `My_Fraction1` is a subclass of the class `handle`, which is a superclass within MATLAB. This allows the use of the methods of this class without having to write them. Thus, it is said that `My_Fraction1` is a subclass or derived class of the superclass `handle`. To obtain information about the superclass `handle` in MATLAB, we need to write `handle` and all the features of this superclass and methods associated with it will be associated (see Figure 7.2).

7.3.2 Declaration and Use of Setters

Due to encapsulation, the only way to modify the contents of the properties or attributes of the class is through the methods declared in them. To modify the contents, functions can be added to set the values of the numerator and denominator. This is done with the following methods:

```
function setNumerator(obj, numen)
   obj.numerator = numen;
end

function setDenominator(obj, denom)
   obj.denominator = denom;
end
```

The class then is (the name has been changed to My_Fraction2):

```
classdef My_Fraction2 < handle
   % This class evaluates the ratio of two numbers.

   properties
      numerator = [ ]
      denominator = [ ]
   end

   methods
      function setNumerator(obj, numen)
         obj.numerator = numen;
      end
      function setDenominator(obj, denom)
         obj.denominator = denom;
      end
      function Result = Fraction(obj)
         Result = obj.numerator/obj.denominator;
      end
   end
end
```

The functions setNumerator and setDenominator are known as **setters** since they allow the initialization (set) of variables. The initialization of the values of the numerator and denominator can be done in either way as shown in the following example for the setting of numerator and denominator. Thus, we have:

```
>> a = My_Fraction2;
>> a.setNumerator(10);
>> a.denominator = 15;
>> frac = a.Fraction

   frac =
       0.6667
```

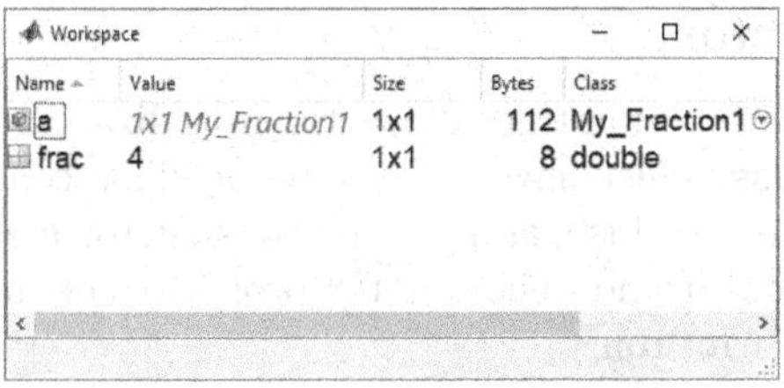

FIGURE 7.3: Characteristics of the object **a** and the variable **frac**.

In the **Workspace** window we can see the variables and the type of the variables. The variable **frac** is of type **double** while the object is an instance of the **class My_Fraction2** as shown in Figure 7.3. Now we make another instance of the class as follows:

```
>> b = My_Fraction2;
>> b.setNumerator(1)
>> b.setDenominator(4)
>> result = b.Fraction
   result =
       0.2500
```

Now object **b** is another instance of the class **My_Fraction2** and the values of its properties are different and independent of those belonging to object **a**.

7.3.3   Inheritance

The properties of classes can be inherited from existing classes. Thus, the class that inherits it is called a subclass. A top-level class is called a superclass. Each subclass can alter the properties which were inherited and add also their own.

In the last definition of the class of the example, the class **My_Fraction2** is a subclass of the superclass **handle** which has its own methods and that is predefined in MATLAB. The class **handle** is an abstract class, which means that one cannot directly create instances of this class. This class is used as a superclass when we implement our own class. The class **handle** is the foundation of all classes.

As the class **My_Fraction2** is a subclass of the class **handle**, it inherits the methods of this class and allows the use of them without having to write them, since it is possible at any time to invoke them whenever the class **My_Fraction2** is used. All classes that follow the semantics of the class **handle**, are subclasses of the class **handle**.

7.3.4   Constructor

The constructor is a function that allows the initialization of the properties of an object. Every class must have a constructor. This constructor must have the same class name. In this example, the constructor method assigns initial values to the numerator and denominator properties of the instances created from the class My_Fraction.

In such a way that the class, which is now called My_Fraction3 contains a declaration of three methods: two methods of type setter (function setNumerator and function setDenominator), a specific method regarding the behavior of the fraction (function Result) and the method constructor (My_Fraction3). The class My_Fraction3 is now:

```
classdef My_Fraction3 < handle
    % This class evaluates the ratio of two numbers.
    properties
        numerator = [ ]
        denominator = [ ]
    end

    methods
        function setNumerator(obj, numen)
            obj.numerator = numen;
        end
        function setDenominator(obj, denom)
            obj.denominator = denom;
        end
        function Result = Fraction(obj)
            Result = obj.numerator/obj.denominator;
        end

    % This is the constructor - - - - - - - - - - - -
        function obj = My_Fraction3 (num, den)
            obj.setNumerator(num);
            obj.setDenominator(den);
        end
    % - - - - - - - - - - - - - - - - - - - - - - - -

    end
end
```

Now we show the creation of an instance of the class My_Fraction3. Due to the constructor, when doing an instance of the class, values must be assigned to the numerator and denominator:

```
>> a = My_Fraction3(8, 9)
   a =
   My_Fraction3 with properties:
      numerator: 8
      denominator: 9
>> result = a.Fraction
   result =
      0.88892500

>> a.setDenominator(4)
>> result = a.Fraction

   result =
      2
```

7.3.5   Direct and Indirect Access to Properties

The properties of a class are accessible from the outside in the same way as the fields of a structure (**struct**). As it has been illustrated, it is possible to access them and assign values using dot notation.

However, respecting the principle of encapsulation and facilitating the development of applications, it is desirable to restrict direct access to the properties to maintain a separation between the implementation and the interface. In this way, access to the class is only possible through methods defined specifically for it. These methods are known as getters and setters.

MATLAB provides mechanisms to keep control on the access to the properties of classes. There are 3 levels: private, protected, and public. They are applied independently for the access to reading and / or writing. Private properties are only accessible from the class methods and public properties (which are the default) are accessible from anywhere. It is possible to specify different access rights for each property, making some public and others private. Protected properties and private properties are also accessible from the subclasses. To mark a group of properties as private, the following statement is used:

```
properties (SetAccess = private)
```

In this case it is no longer possible to define any of the two properties from the outside as

```
a.numerator = 10
```

in which case it will indicate an error because there is an indirect access. The next message is then shown:

> You cannot set the read-only property 'numerator' of
> My_Fraction3.

Having declared that the access to the properties is private, it is only possible to manipulate them with the setters method. This is called a direct access. When the properties are private and it is desired to add other features to the object, it is necessary to modify the corresponding setters method. For example, to perform validations and prevent the denominator in the fraction to be zero, the setter must be modified as follows:

```
function setDenominator(obj, denom)
   if denom == 0
      error('The denominator must be different from 0.')
   end

   obj.denominator = denom;
end
```

And if it is wanted to create an instance where the denominator is equal to 0, the following error message is shown:

```
>> b = My_Fraction3(3, 0)
   Error using My_Fraction3/setDenominator (line 14)
   The denominator must be different from 0.
   Error in My_Fraction3 (line 14)
      obj.setDenominator(den);
```

Additionally, the type of variables can be checked. As in the example, the intention is to form a fraction, where the values of both the numerator and denominator should only be integer, i.e., that they are not strings or that have a fractional part. This can be done if the next condition applies in the setter of the numerator :

```
if mod(numerator, 1) ~= 0 || isstring(numerator) == 1
   error('The numerator must be an integer number.')
end
```

Adding the same condition to the if of the denominator setter:

```
function setDenominator(obj, denom)
   if denom == 0 || mod(denom, 1) ~= 0 || isstring(denom) == 1
      error('Denominator must be different from 0.')
   end
   obj.denominator = denom;
end
```

With these changes, the class, now referred to as My_Fraction4 is:

```
classdef My_Fraction4 < handle
    % This class evaluates the ratio of two numbers.

    properties (SetAccess = private)
        numerator = [ ]
        denominator = [ ]
    end

    methods
        function setNumerator(obj, numen)
            if mod(numen, 1) ~= 0 || ischar(numen) == 1
                error('The numerator must be an integer number.')
            end
            obj.numerator = numen;
        end
        function setDenominator(obj, denom)
            if denom == 0 || mod(denom, 1) ~= 0 || ischar(denom)==1
                error('Denominator must be different from 0.')
            end
            obj.denominator = denom;
        end
        function result = Fraction(obj)
            result = obj.numerator/obj.denominator;
        end

        % This is the constructor - - - - - - - - - - - - - - - - -

        function obj = My_Fraction4 (num, den)
            obj.setNumerator(num);
            obj.setDenominator(den);
        end
    % - - - - - - - - - - - - - - - - - - - - - - - - - - - - - - -
    end
end
```

Let us now illustrate the creation of an instance of the class My_Fraction4, in which the denominator must be different from 0:

```
>> a = My_Fraction4(3, 0)
    Error using My_Fraction4/setDenominator (line 17)
    The denominator must be integer different from 0.

    Error in My_Fraction4 (line 27)
    obj.setDenominator(den);
```

```
>> a = My_Fraction4(3, 4);
>> fracc = a.Fraction
   fracc =
       0.7500
```

7.3.6 Public and Private Methods

Sometimes it is desirable or necessary to have methods that are used only within the same class. This means that they cannot be invoked by any other class. For example, in the My_Fraction class that has been defined, when forming the fraction, it would be desirable to form it with minimum factors, i.e. the minimum values for the numerator and denominator. To achieve this, it is needed to calculate the greatest common divisor (gcd) to reduce both the numerator and the denominator and simplified values, forming the fraction. It is not necessary that the method for reduction is visible to the other classes since it is only used by other methods of the same class. For the greatest common divisor MATLAB has the function gcd(a, b) that calculates the gcd of a and b. Let us see how the method reduction is declared in the following code:

```
methods (Hidden = true)

    function reduce(obj)
    % First we check that we have input data.
        if isempty(obj.numerator) || isempty(obj.denominator)
            return
        end

    % Check if anyone is negative.
        is_negative = obj.numerator*obj.denominator < 0;

    % We obtain the GCD of the absolute values.
        mygcd=gcd(abs(obj.numerator),abs(obj.denominator));

    % We reduce the numbers.
        obj.numerator = abs(obj.numerator)/mygcd;
        obj.denominator = abs(obj.denominator)/mygcd;

    % If any of the numbers is negative,
    % change the sign of the numerator.
        if is_negative
            obj.numerator = -obj.numerator;
        end
    end
end
```

In MATLAB, methods are declared as private, marking as true the parameter `Hidden`. The instruction `Hidden = true`, ensures that this method is hidden and cannot be used from outside the class.

Once the method **reduce** is defined, it can be used by any other method of the same class. For example, it may be added to the methods `setDenominator` and `setNumerator` by adding the line:

$$\texttt{obj.reduce()}$$

to perform the simplification of the numbers. The method `setNumerator` is then

```
function setNumerator(obj, numen)
    if mod(numen, 1) ~= 1 || ischar(numen) == 1
        error ('The numerator must be an integer.')
        obj.numerator = numen;
        obj.reduce()

    end

end
```

Similarly, the method **reduce** is added to the method `setDenominator`. The complete file is saved as `My_Fraction5` and is:

```
classdef My_Fraction5 < handle
    % This class evaluates the ratio of two numbers.

    properties (SetAccess = private)
        numerator = [ ]
        denominator = [ ]
    end

    methods (Hidden = true)

        function reduce(obj)
        % First we check that we have input data.
            if isempty(obj.numerator) || isempty(obj.denominator)
                return
            end

        % Check if anyone is negative.
            is_negative = obj.numerator*obj.denominator < 0;

        % We obtain the GCD of the absolute values.
```

```
          mygcd=gcd(abs(obj.numerator), abs(obj.denominator));

   % We reduce the numbers.
          obj.numerator = abs(obj.numerator)/mygcd;
          obj.denominator = abs(obj.denominator)/mygcd;

   % If any of the numbers is negative,
   % change the sign of numerator.
          if is_negative
              obj.numerator = -obj.numerator;
          end
      end
   end

   methods
      function setNumerator(obj, numen)
         if mod(numen, 1) ~= 0 || ischar(numen) == 1
            error('The numerator must be an integer number.')
         end
         obj.numerator = numen;
         obj.reduce;
      end
      function setDenominator(obj, denom)
         if denom == 0 || mod(denom, 1) ~= 0 || ischar(denom)==1
            error('Denominator must be different from 0.')
         end
         obj.denominator = denom;
         obj.reduce;
      end
      function result = Fraction(obj)
         result = obj.numerator/obj.denominator;
      end

   % This is the constructor - - - - - - - - - - - - - - - -
      function obj = My_Fraction5 (num, den)
         obj.setNumerator(num);
         obj.setDenominator(den);
      end
   % - - - - - - - - - - - - - - - - - - - - - - - - - - - - -
   end
end
```

Now we run the class My_Fraction5. First, we make the denominator a string:

```
>> a = My_Fraction5(4, '12')
```

```
Error using My_Fraction5/setDenominator (line 37)
The denominator must be different from 0.
Error in My_Fraction5 (line 47)
   obj.setDenominator(den);
```

Now we make the numerator a string:

```
>> a = My_Fraction5('4', 12)
```

```
Error using My_Fraction5/setNumerator (line 13)
The numerator must be an integer.
Error in My_Fraction5 (line 38)
   obj.setNumerator(num);
```

Finally, valid values for the numerator and the denominator are given:

```
>> a = My_Fraction5(4, 12)
a =
My_Fraction5 with properties:
numerator: 1
denominator: 3
>> a.Fraction
ans =
0.3333
```

In this case, the values of the numerator and denominator have already been simplified and reduced by dividing the greatest common divisor of both numbers which in this case is `gcd = 4`.

7.3.7 Overriding Methods

The methods that a class inherits from the superclass can be used by it because it inherits all the methods defined in the superclass, along with those defined in itself. In addition, the class can redefine a method of the superclass, creating a function with the same signature (name, input arguments and output), so that the methods of the superclass are redefined or replaced. This is known as overriding methods.

In this example, we said at the beginning that given two integers, we wanted to form a fraction. However, it has been seen that fractions are being calculated and the result of the fraction was delivered showing the ratio of the numerator over the denominator. One way to avoid this is to redefine the way to display the result, showing the fraction without evaluating it. The class that has been defined, by being a subclass of the superclass `handle`, inherits the behavior of the same and therefore uses the method `disp`. But in the new class, the behavior of method `disp` can be rewritten to accommodate how the fraction is to be displayed. This is done by adding the following method:

```
function disp(obj)
    fprintf(' %d / %d \n \n', obj.numerator, obj.denominator)
end
```

We add this method to class `My_Fraction5` and rename the resulting class as `My_Fraction6`.

```
classdef My_Fraction6 < handle
    % This class evaluates the ratio of two numbers.
    properties (SetAccess = private)
        numerator = [ ]
        denominator = [ ]
    end
    methods (Hidden = true)
        function reduce(obj)
        % First we check that we have input data.
            if isempty(obj.numerator) || isempty(obj.denominator)
                return
            end
        % Check if anyone is negative.
            is_negative = obj.numerator*obj.denominator < 0;
        % We obtain the GCD of the absolute values.
            mygcd=gcd(abs(obj.numerator), abs(obj.denominator));
        % We reduce the numbers.
            obj.numerator = abs(obj.numerator)/mygcd;
            obj.denominator = abs(obj.denominator)/mygcd;
        % If any of the numbers is negative,
        % change the sign of numerator.
            if is_negative
                obj.numerator = -obj.numerator;
            end
        end
    end
    methods
        function setNumerator(obj, numen)
            if mod(numen, 1) ~= 0 || ischar(numen) == 1
                error('The numerator must be an integer number.')
            end
            obj.numerator = numen;
            obj.reduce;
        end
```

```
function setDenominator(obj, denom)
    if denom == 0 || mod(denom, 1) ~= 0 || ischar(denom)==1
        error('Denominator must be different from 0.')
    end
    obj.denominator = denom;
    obj.reduce;
end
function result = Fraction(obj)
    result = obj.numerator/obj.denominator;
end
function obj = My_Fraction6 (num, den)
    obj.setNumerator(num);
    obj.setDenominator(den);
end
% - - - - - - - - - - - - - - - - - - - - - - - - - - - - - - - -
function disp(obj)
    fprintf('%d/%d \n\n',obj.numerator,obj.denominator)
end
% - - - - - - - - - - - - - - - - - - - - - - - - - - - - - - - -
    end
end
```

We run now the following

```
>> a = My_Fraction6(12, 16)
   a =
   3/4
```

We see that the result is given as a fraction after removing the common factors.

7.3.8   Overloading

Overloading is the mechanism that allows a method to accept different types of arguments and to have an existing name. Let us illustrate this mechanism through this example by incorporating a method that allows the addition of fractions. Let us see what happens when trying to add the objects a and b:

```
>> a + b
   Undefined operator '+' for input arguments of type
   'My_Fraction6'.
```

MATLAB reports an error, since it has not been defined the operation that allows the addition of two objects of the type My_Fraction6. Recall that for the basic arithmetic operations of addition (+), subtraction (-), multiplication (*) and division (/), MATLAB establishes the equivalence with the following functions: `indexplus`, `minus`, `mrtimes` and `mrdivide`. The information that the MATLAB help provides about the function `plus` is obtained as:

```
>> help plus

   + Plus
   X + Y adds matrices X and Y. X and Y must have the same
   dimensions unless one is a scalar (a 1-by-1 matrix).
   A scalar can be added to anything.

   C = plus(A,B) is called for the syntax 'A + B' when A
   or B is an object.

Reference page for plus
```

Since MATLAB does not offer what is needed to make the sum of fractions, then a new method can be developed, suitable for short fractions. Consider the addition of two fractions as follows:

$$\frac{A}{B} + \frac{C}{D} = \frac{AD + BC}{BD}$$

for which the following method is declared:

```
function resulta = plus(obj, obj2)
    resulta = My_Fraction(obj.numerator*obj2.denominator + ...)
       obj.denominator*obj2.numerator, ...
       obj.denominator*obj2.denominator);
end
```

In the same way, a method can be implemented for the subtraction of the two fractions by rewriting the function `uminus`, which will change the sign of the fraction:

```
function resultu = uminus(obj)
    resultu = My_Fraction7(-obj.numerator, obj.denominator);
end
```

and redefining afterwards the function `minus` to perform the subtraction as an addition with a negative number:

```
function resultm = minus(obj, obj2)
    resultm = obj.plus(-obj2);
end
```

By incorporating these definitions of new methods to perform arithmetic operations on the fractions, a new class called My_Fraction7 is created.

```
classdef My_Fraction7 < handle
    % This class evaluates the ratio of two numbers.

    properties (SetAccess = private)
        numerator = [ ]
        denominator = [ ]
    end

    methods (Hidden = true)

        function reduce(obj)
        % First we check that we have input data.
            if isempty(obj.numerator) || isempty(obj.denominator)
                return
            end

        % Check if anyone is negative.
            is_negative = obj.numerator*obj.denominator < 0;

        % We obtain the GCD of the absolute values.
            mygcd=gcd(abs(obj.numerator), abs(obj.denominator));

        % We reduce the numbers.
            obj.numerator = abs(obj.numerator)/mygcd;
            obj.denominator = abs(obj.denominator)/mygcd;

        % If any of the numbers is negative,
        % change the sign of numerator.
            if is_negative
                obj.numerator = -obj.numerator;
            end
        end
    end

    methods

        function setNumerator(obj, numen)
            if mod(numen, 1) ~= 0 || ischar(numen) == 1
                error('The numerator must be an integer number.')
            end
            obj.numerator = numen;
            obj.reduce;
        end
```

```matlab
    function setDenominator(obj, denom)
        if denom == 0 || mod(denom, 1) ~= 0 || ischar(denom)==1
            error('Denominator must be different from 0.')
        end
        obj.denominator = denom;
        obj.reduce( );
    end
    function result = Fraction(obj)
        result = obj.numerator/obj.denominator;
    end
    function obj = My_Fraction7(num, den)
        obj.setNumerator(num);
        obj.setDenominator(den);
    end
    function disp(obj)
        fprintf('%d/%d \n\n',obj.numerator,obj.denominator)
    end


% - - - - - - - - - - - - - - - - - - - - - - - - - - - -
function resulta = plus(obj, obj2)
    resulta = My_Fraction7(obj.numerator*obj2.denominator + ...
        obj.denominator*obj2.numerator, ...
        obj.denominator*obj2.denominator);
end
function resultu = uminus(obj)
    resultu = My_Fraction7(-obj.numerator, obj.denominator);
end

function resultm = minus(obj, obj2)
    resultm = obj.plus(-obj2);
end

function resultp = mtimes(obj, obj2)
    resultp = My_Fraction7(obj.numerator*obj2.numerator,...
        obj.denominator*obj2.denominator);
end

function resultd = mrdivide(obj, obj2)
    resultd = My_Fraction7(obj.numerator*obj2.denominator,...
        obj.denominator*obj2.numerator);
end
% - - - - - - - - - - - - - - - - - - - - - - - - - - - -
    end
end
```

Two instances of `My_Fraction7` are created to perform the addition and subtraction operations, as shown below:

```
>> a = My_Fraction7(3, 9)
   a =
   1/3
>> b = My_Fraction7(2, 14)
   b =
   1/7

>> a + b
   ans =
   10/21
```

Thus, it is observed that the result is correct. Testing for the subtraction:

```
>> a - b
   ans =
   4/21
```

as expected. For the multiplication, MATLAB uses the function `mtimes` which can be overwritten as:

```
function resultp = mtimes(obj, obj2)
    resultp = My_Fraction7(obj.numerator*obj2.numerator, ...)
        obj.denominator*obj2.denominator);
end
```

and for the division:

```
function resultd = mrdivide(obj, obj2)
    resultd = My_Fraction7(obj.numerator*obj2.denominator, ...)
        (obj.denominator*obj2.numerator);
end
```

Now fractions a and b can be multiplied and divided:

```
>> a*b
   ans =
   1/21

>> a/b
   ans =
   7/3
```

7.4   Examples

In this section two examples of classes are shown that illustrate the main features of MATLAB for applying the object-oriented paradigm.

Example 7.1 The opening and movements of a bank account.

This example shows the way to implement the movements associated with a bank account, using object orientation offered by MATLAB. The movements that must be performed are opening the account, deposits, withdrawals and the displaying of the balance. The main characteristic of any account is the balance, thus it is necessary to specify methods for making deposits, withdrawals and obtaining the balance. After each deposit and / or withdrawal it is also required to see the balance, for which a special method to display it will be used.

Let us begin with the declaration of the class called `Bank_Account` whose basic structure is shown in the following listing:

```
classdef Bank_Account < handle
    % It creates and manages accounts.

    properties
        balance = [ ]
    end
    methods
    end
end
```

The first method is the constructor that must bear the same name as the class which will create the account and assign a value to the opening balance. Each object account will then be an object or an instance of the class `Bank_Account`. The constructor must read the account name and the value to be assigned to the opening balance. An appropriate constructor is:

```
% Constructor
function account = Bank_Account(data)
    % Constructor of the class cb = Bank_Account.
    % Create a bank account with a balance 'data'.

    if nargin == 0 % This option occurs when no data is given.
        account.balance = 0;
    else
        % This option is to initialize the balance.
        account.balance = data;
```

```
        end
    end
```

Now the methods to deposit and withdraw are written. In the first case it is an addition of the amount received to the previous balance, and in the second case it is subtracted. Also for consideration is the case where the withdrawal is greater than the balance. In both methods, the account represents an instance of the class **Bank_Account** to modify and the amount represents the amount to add or subtract from the balance of the account. For the deposit, the method to define is:

```
function deposit(account, amount)
    % Increments the balance by amount.
    compute_balance(account, account.balance + amount);
end
```

and for the withdrawal the method is:

```
function amount_withdraw = withdrawal(account, amount)
    % Decrease account balance by the amount.
    amount_withdraw = amount;
    if amount_withdraw > account.balance
        sprintf('Insufficient funds.')
    else
        compute_balance(account, account.balance...
            - amount_withdraw);
    end
end
```

For the balance, now the following method is defined:

```
function compute_balance(account, amount)
    % Obtain the balance of the account.
    account.balance = amount;
end
```

Finally, to display the results the following methods are used:

```
function display(cb)
    % Displays the representation of the bank account
    % as a string.
    %
    disp(char(cb));
end
```

```
function s = char(cb)
   % Bank_Account.
   % Representation as a string.
   s = sprintf('account with $ % .2f \n', cb.balance );
end
```

A method to calculate the balance of any of the objects can be added. This is done with:

```
function obtain_bal = obt_balance(cb)
   % obtains the balance
    obtain_bal = cb.balance;
    display(char(cb))
end
```

To test how it works, two objects named Peter and John are defined:

```
>> Peter = Bank_Account(12)
   account with $12.00

>> John = Bank_Account(45)
   account with $45.00

>> John.deposit(14);

>> compute_balance(John)
   account with $59.00
   ans =
   59

>> Peter.withdrawal(89)
   ans =
   Insufficient funds.

>> Peter.withdrawal(17)
   ans =
   Insufficient funds.

>> Peter.withdrawal(7)

>> Peter.balance
   ans =
   5

>> compute_balance(Peter)
   account with $5.00
   ans =
   5
```

Example 7.2 Automatic differentiation.

The subject of automatic differentiation arises from the need for greater accuracy and speed in the calculation of derivatives. This is because a symbolic differentiation is slower and a numerical differentiation has rounding errors. In automatic differentiation it is necessary to give the value of the function and value of its derivative. The function is partitioned into its basic components, differentiation is made on them and finally the function is formed. For example, the function $f(x) = x^2$ decomposes into the function $f(x) = x*x$. As another example, the function $f(x) = (x + x^2)^2$ decomposes as:

```
z = x*x          x∧2
w = x + z        x + x∧2
y = w*w          (x + x∧2)∧2
```

To write the class in OOP code, first a new class is created with the name differentiator. The properties of this class are the input data that is the value of the function and the value of its derivative. The constructor reads the values of the properties. The initial arrangement of the class is as shown below:

```
classdef differentiator < handle
  % differentiator: Implements automatic differentiation.
  % Obtains the first derivative using simple operations.
  % Needs a value for the function and
  % a value for the derivative.
  properties
      value % value of the function.
      derivative % value of the derivative.
  end
  methods
      function obj = differentiator(a, b) % Constructor.
          if nargin == 0 % No data given. Do nothing.
              obj.value= [ ];
              obj.derivative = [ ];
          elseif nargin == 1% c = differentiator(a)
                            % for derivative = 0.
              obj.value = a;
              obj.derivative = 0;
          else
              obj.value = a;  % value of the function.
              obj.derivative = b; % value of the derivative.
          end
      end
  end
end
```

Now the methods which are going to perform automatic differentiation are written. For example, for the sine function, it originates from the traditional formula of the derivative of the sine that is:

$$\frac{d\,\sin[f(x)]}{dx} = \cos[f(x)] * \frac{df(x)}{dx}$$

then we can write the method for the sine derivative as:

```
function h = sin(u)
  % This definition overloads the function sin from MATLAB.
  h = differentiator(sin(u.value), u.derivative*cos(u.value));
end
```

Similarly, the methods to the four basic operations, for trigonometric functions, hyperbolic functions, transcendental functions, etc., can overloaded. To illustrate this, a multiplication is overloaded in MATLAB with mtimes. Here the instruction isa is used, which checks whether an object is of the corresponding class. The corresponding format is:

```
isa(object, 'name of the class')
```

The method mtimes is then:

```
function hh = mtimes(u, v)

  % Here, we overload the multiplication.
  % One of the factors may not be an object
  % of the class differentiator.

  if ~isa(u, 'differentiator') % Checks if u is a scalar.
     hh = differentiator(u*v.value, u*v.derivative);
  elseif ~isa(v, 'differentiator') % Checks if v is a scalar.
     hh = differentiator(v*u.value, v*u.derivative);
  else
     hh = differentiator(u.value*v.value,...% Both are objects.
        u.derivative*v.value + u.value*v.derivative);
  end

end
```

Other methods are on the book's website and can be downloaded to run the class. As an example, to obtain the value of the first derivative of the function sin(x*x) at x = $\pi/4$:

```
>>  x = differentiator(pi/4, 1)
    x =
    differentiator with properties:

    value: 0.7854
    derivative: 1

>>  sin(x*x)
    ans =
    differentiator with properties:

    value: 0.5785
    derivative: 1.2813
```

which indicates that the value of the function sin(x*x) at $\pi/4$ is 0.5785 and the value of the first derivative of this function is 1.2813.

7.5 Conclusions

In this chapter, an introduction to using MATLAB to write programs using the paradigm of Object-Oriented Programming was given. Only the most basic definitions have been explored since the topic is quite large and is beyond the aims of this book and chapter. The interested reader can consult the references for a greater knowledge of the subject. However, with this chapter it is already possible to create classes of OOP in MATLAB.

Chapter 8

Graphical User Interfaces

A graphical user interface (GUI) is the link between a software package and the user. In general, it is composed of a set of commands or menus, objects and instruments such as buttons, by means of which the user establishes a communication with the program. The GUI eases the tasks of inputting data and displaying output data.

8.1   Creation of a GUI with the Tool GUIDE

MATLAB has a tool to develop GUIs in an easy and quick way. This tool is called **Graphical User Interface Development Environment** and is better known by its acronym GUIDE. This tool can create a GUI empty window, add buttons and menus to our GUI, and windows to enter data and plot functions, as well as the access to the objects callbacks. When we create a GUI with GUIDE, two files are created: a fig-file which is the graphical interface and an m-file which contains the functions, the description for the GUI parts, and the callback.

A **callback** is defined as the action that implements an object of the GUI when the user clicks on it or uses it. For example, when the user clicks on a button in a GUI, a program containing the instructions and tasks to be realized is executed. This program is called the callback. A callback is coded in the m-language.

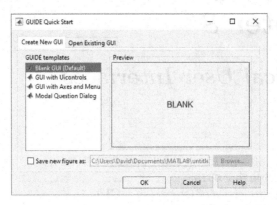

FIGURE 8.1: `GUIDE` Quick Start window.

8.1.1 Starting `GUIDE`

`GUIDE` can be started from the MATLAB menu with `New→ App→ GUIDE`. It can also be started by typing `guide` in the `Command Window`. Any of these choices opens the `GUIDE Quick Start` window shown in Figure 8.1. Here we can start a new GUI with an empty blank GUI where we can add and arrange the object for the GUI. We can also start a new GUI with `uicontrols`, with a set of axes and a menu already in the GUI, and finally, a GUI with question dialog buttons. Alternatively, we can open a GUI previously started in the tab `Open Existing GUI`. If we select `Blank GUI` we get the work window of Figure 8.2. In this work window we see at the left a set of buttons or objects that can be used in the GUI. (To see the object names in the buttons, in the `File` menu select `File → Preferences ... → GUIDE` and select the option `Show names in component palette`.) Each button has a function which is described by the button's name and it is self describing.

On the upper part we see the toolbar. It contains icons to create a new GUI or figure, to open an existing GUI, and to save the GUI. It also has icons to copy, paste, and cut parts of the GUI, as well as to undo and redo actions on the GUI objects. In addition, the toolbar has icons to align the objects in the GUI, another icon for the `Editor` and for the `Property Inspector`, to display the browser, and to execute the GUI. Some properties are further explained in Table 8.1.

8.1.2 Properties of Objects in a GUI

Each object that we place in the GUI has properties that can be edited with the `Property Inspector`. For example, for a `Push Button` Figure 8.3 shows the `Property Inspector` with some of the properties of this button. Some of the most common properties are shown in Table 8.2.

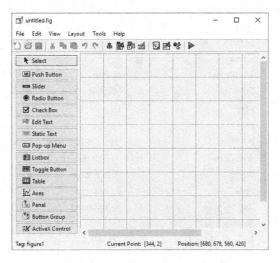

FIGURE 8.2: A blank GUI.

TABLE 8.1: Important icons in the `GUIDE` toolbar.

Property icon name	Description
Property inspector	It refers to the properties of each object in the GUI. They include color, name, tag, value, and the callback among others.
Align Objects	It aligns the objects in the work window.
Toolbar editor	It creates a toolbar in the GUI.
M-file editor	It opens the MATLAB editor to edit the callbacks.

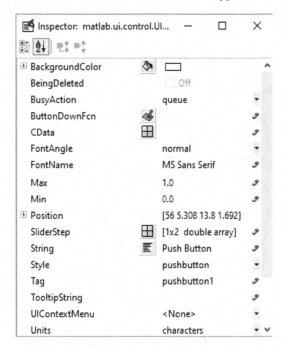

FIGURE 8.3: Property inspector.

8.1.3   A Simple GUI

We show with a plotting example the procedure to create a GUI. Let us suppose that we wish to create a GUI that plots a user defined function. Thus, we need a text box to enter the function, two text boxes to enter the initial and final points in the plot, a set of axes to plot the function, and a button to run the GUI. Additionally, we can add a button to close the GUI after we finish.

1. The first step is to open a blank GUI work window.

2. We add the required objects.

   - We start with two push buttons. A button to plot the function and another one to close the GUI.

   - Three `Edit Text` boxes, a text box to enter the function to be plotted and two text boxes for the x-axis limits.

   - A set of axes.

   - Five `Static Texts` for labels. The GUI is shown in Figure 8.4.

TABLE 8.2: Most used properties for objects in a GUI.

Property	Description
Background color	Changes the background color of the object.
Callback	Set of instructions to be executed by the object.
Enable	Activates the object.
String	Mostly used in the cases of buttons, edit text boxes, and static text boxes. It contains the text displayed in the object.
Tag	It identifies the object.

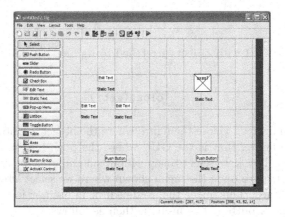

FIGURE 8.4: GUI with required objects.

3. We can stretch each of the objects to the desired final size. We also change the `String` property of each object as we can see in Figure 8.5. To change the `String` property, we double click on each one of the objects to open the `Property inspector` and change the `String` in the `Static texts`. The `Font weight` property is changed to bold. For each of the remaining elements we clear the `Strings`.

4. For the `Edit Text` box we change the `Tag` property to `The_function`. This is the variable name for the function to be plotted. For the remaining `Edit Text` boxes we change the tags to `Initial_x` and `Final_x`.

5. We change the `Tag` properties of the push buttons to `Plot_function` and `CloseGUI`.

6. We save the GUI as `plotter.fig`.

7. We now need to edit the callbacks.

8. First we edit the callback of the `Close` button. We only need to add the instruction:

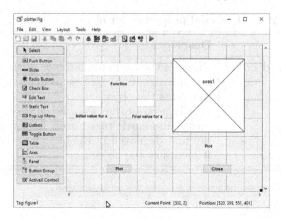

FIGURE 8.5: GUI with objects stretched to its final size and final labels.

```
close(gcbf)
```

which indicates to close the figure where the object is embedded. In this case the figure, that is the GUI, where the push button is located. To edit the callback we select this push button and click on the mouse right hand button to open the menu shown where we select View callbacks → Callback. This opens the m-file editor in the portion corresponding to this push button. The m-file is then as follows:

```
% -- Executes on button press in CloseGUI.
function CloseGUI_Callback(hObject, eventdata, handles)
% hObject handle to CloseGUI (see GCBO)
% eventdata reserved - to be defined in a future version of
MATLAB
% handles structure with handles and user data (see GUIDATA)
close(gcbf)
```

9. We now edit the callback for the button Plot. In this callback we have to read the function we wish to plot. We also read the lower and upper limits for the variable x which we declare as a symbolic variable with syms x. Finally, we plot the function in the set of axes in the GUI. When we read data from a GUI, this data is read as a string. Thus, at some point we have to put the data in the correct format, for example an integer variable, a boolean variable, and so on. First, we open the callback for the button Plot by right clicking on the push button and selecting View callbacks → Callback. This opens the m-file editor. To read the information in the strings and make it amenable for calculations we

use the instruction **eval** to convert the string value we have read to a numeric value. For example, for the lower limit of x

```
lower_x_value = eval(get(handles.Initial_x, 'string'));
```

The instruction **get** fetches the string variable which is located in the object with the handle **Initial_x** and the instruction **eval** converts it to a real variable. A similar instruction applies for the upper x limit and for the function to be plotted. That is,

```
lower_x_value = eval( get( handles.Initial_x, 'string'));
upper_x_value = str2num(get( handles.Final_x, 'string'));
y = eval(get( handles.The_function, 'string'));
```

We are using **eval** and **str2num** to convert from string to a number. They both accomplish the same task.

10. Now, we are ready to get the plot. We now get the x-axis points with

```
xx = [lower_x_value:0.2:upper_x_value];
```

11. The function to be plotted is now in the variable y which is a string. To be able to accept x as a symbolic variable, we add the instruction **syms** x. This makes the variable y a symbolic variable.

12. To evaluate the function y at the set of points xx we substitute the variable x with the vector xx with

```
yb = subs(y, x, xx);
```

13. Now, we plot the vector yb with

```
plot(xx, yb)
```

14. Finally, we add a grid with

```
grid on
```

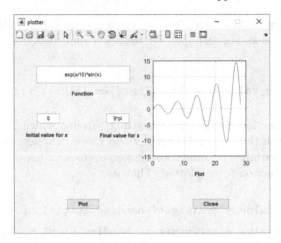

FIGURE 8.6: Plot of function $e^{(x/10)}\sin(x)$ from 0 to 9π.

15. The final callback for the push button `Plot` is

```
% -- Executes on button press in Plot_function.
function Plot_function_Callback(hObject, eventdata, handles)
% hObject handle to Plot_function (see GCBO)
% eventdata reserved - to be defined in a future version of
MATLAB
% handles structure with handles and user data (see GUIDATA)
syms x
lower_x_value = eval( get( handles.Initial_x, 'string'));
upper_x_value = str2num( get( handles.Final_x, 'string'));
xx = [lower_x_value:0.2:upper_x_value];
y = eval(get( handles.The_function, 'string'));
yb = subs(y, x, xx);
plot(xx, yb)
grid on
```

16. To finish the GUI we add a toolbar. From the `Tools` menu select `Toolbar Editor` .... When it opens change the icons as desired. For this example we only click the `OK` button to display the default toolbar. Once we are finished we save the current GUI as `plotter`.

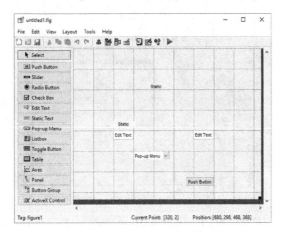

FIGURE 8.7: Initial layout for the GUI.

Note that the complete GUI is composed of the interface that is saved in `plotter.fig`, that is a figure file, and the m-file `plotter.m` that contains the callbacks for the objects in the GUI.

Now, to run the GUI either we type `plotter` in the MATLAB command window or click on the Run icon in the GUI work window. Figure 8.6 shows a plot for the function `exp(x/10)*sin(x)`.

8.2 Examples

This section presents two examples introducing some of the other objects for the GUIs. The first example shows the pulldown menu for a GUI that converts temperature given in Fahrenheit degrees to Celsius and vice versa. The second example is a GUI for the calculation of put and call options using the Black-Scholes function from the Financial Derivatives toolbox.

Example 8.1 Temperature conversion

Temperature conversion among the different scales, Celsius, Fahrenheit, and Kelvin, is possible if we know the conversion equations. They are available in any physics textbook and they are:

```
F = 1.8*C + 32
K = C + 273.15
C = (F - 32)*5/9
K = (F - 32)*5/9 + 273.15
C = K - 273.15
F = 1.8*(K - 273.15) + 32
```

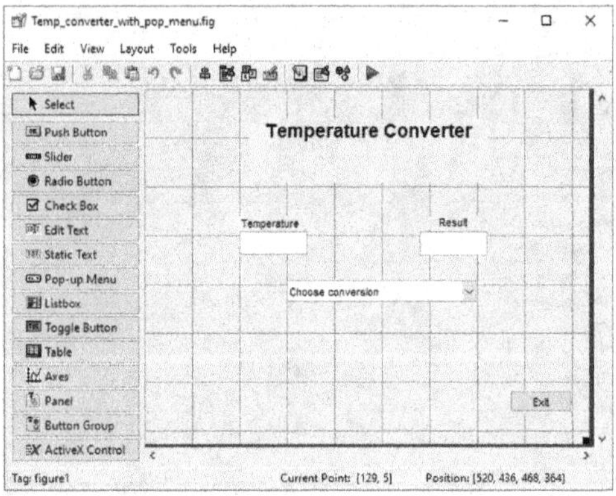

FIGURE 8.8: GUI layout with strings and sizes changed.

FIGURE 8.9: String for the pop-up menu. Click on the icon to the right of
String.

Now we create a GUI that implements these conversion equations. The initial
GUI layout is shown in Figure 8.7. We now implement the instructions for each
object. We use the regular instructions to close the GUI and to read data. To
choose the conversion we use a **Pop-up Menu** and for the input temperature
data we use an **Edit Text** box. We have changed the strings and size of the
objects for each of the GUI components so that they look as shown in Figure
8.8. The **Static Text** to the right of the **Edit text** box for the result has a
blank string. The string for the **Pop-up Menu** is set by clicking on the string
property at the **Property Inspector** for the menu. This opens the **String**
window for the **Pop-up Menu** where we add the information shown in Figure
8.9. We now change the tags for each of the components, as shown in Table
8.3.

TABLE 8.3: Tag names.

GUI Component	Tag
Pop-up menu	temp_conv
Edit box below the text Temperature	input_temp
Edit box below the text Result	result
Push button	closeGUI
Static text for Result	degrees

We save the GUI with the name `Temp_converter.m`. We design the GUI in such a way that when the user chooses the conversion with the Pop-up Menu, the conversion takes place and the result is written in the result box. The callback we need to edit is the one corresponding to the Pop-up Menu. First we need to read in the temperature from the Edit Text box. We do this with

```
temp = eval(get(handles.input_temp, 'string'))
```

Note that the value of temperature stored in the Edit Text box is stored in the variable `temp`. Now, we add the instructions to read the conversion from the Pop-up Menu. The variable `val` indicates which conversion we implement with:

```
val = get(hObject, 'Value');
switch val
   case 2
      % Celsius to Fahrenheit
      resultt = temp*1.8 + 32;
      set(handles.degrees, 'string', 'Fahrenheit')
   case 3
      % Celsius to Kelvin
      resultt = temp + 273.15;
      set(handles.degrees, 'string', 'Kelvin')
   case 4
      % Fahrenheit to Celsius
      resultt = (temp - 32)*5/9;
      set(handles.degrees, 'string', 'Celsius')
   case 5
      % Fahrenheit to Kelvin
      resultt = (temp - 32)*5/9 + 273.15;
      set(handles.degrees, 'string', 'Kelvin')
   case 6
      % Kelvin to Celsius
      resultt = temp - 273.15;
      set(handles.degrees, 'string', 'Celsius')
```

```
    case 7
       % Kelvin to Fahrenheit
       resultt = (temp − 273.15)*1.8 + 32;
       set(handles.degrees, 'string', 'Fahrenheit')
end
```

Finally, we write the result to the edit text box result:

```
set(handles.result, 'string', resultt)
```

The complete callback is listed now:

```
% Executes on selection change in temp_conv.
%
function temp_conv_Callback(hObject, eventdata, handles)
% hObject handle to temp_conv (see GCBO).
% handles structure with handles and user data (see GUIDATA).
% Hints: contents = get(hObject, 'String') returns temp_conv
%   contents as cell array.
% contents get(hObject,'Value') returns selected item
% from temp_conv.
temp = eval(get(handles.input_temp,'string'));
val = get(hObject, 'Value');

switch val
   case 2
      % Celsius to Fahrenheit
      resultt = temp*1.8 + 32;
      set(handles.degrees, 'string', 'Fahrenheit')
   case 3
      % Celsius to Kelvin
      resultt = temp + 273.15;
      set(handles.degrees, 'string', 'Kelvin')
   case 4
      % Fahrenheit to Celsius
      resultt = (temp − 32)*5/9;
      set(handles.degrees, 'string', 'Celsius')
   case 5
      % Fahrenheit to Kelvin
      resultt = (temp − 32)*5/9 + 273.15;
      set(handles.degrees, 'string', 'Kelvin')
   case 6
      % Kelvin to Celsius
      resultt = temp − 273.15;
      set(handles.degrees, 'string', 'Celsius')
```

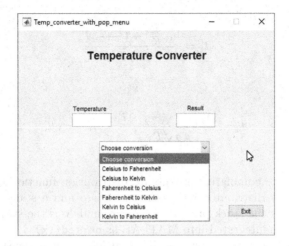

FIGURE 8.10: A run for the temperature conversion GUI.

```
case 7
    % Kelvin to Fahrenheit
    resultt = (temp − 273.15)*1.8 + 32;
end
set(handles.degrees, 'string', 'Fahrenheit')
```

and a run is shown in Figure 8.10.

Example 8.2 Solution of the Black-Scholes equation.
In Chapter 12 we show how to calculate call and put options for European options. There we show that we have to solve the Black-Scholes differential equation:

$$\frac{\partial f}{\partial t} + rS\frac{\partial f}{\partial S} + \frac{1}{2}\sigma^2 S^2 \frac{\partial^2 f}{\partial S^2} = rf$$

whose solution is given by:

1. For the call option

$$c = S_0 N(d_1) K e^{-rT} N(d_2)$$

2. For the put option

$$p = K e^{-rT} N(-d_2) - S_0 N(-d_1)$$

where

$$d_1 = \frac{\ln(\frac{S_0}{K}) + (\frac{r+\sigma^2}{2})T}{\sigma\sqrt{T}}$$

$$d_2 = \frac{\ln(\frac{S_0}{K}) + (\frac{r-\sigma^2}{2})T}{\sigma\sqrt{T}} = d_1 - \sigma\sqrt{T}$$

Here $N(x)$ is the cumulative probability distribution function for a variable that is normally distributed with a mean of zero and a standard deviation equal to 1, S_0 is the stock price at time zero, and K is the strike price. The function $N(x)$ is integrated into MATLAB as normcdf(x).

In this example, we construct a GUI that has as input the stock price S_0, the strike price K, the maturity time T, the interest rate variation r, and the volatility σ. We use the Black-Scholes function from the Financial Derivatives toolbox that has the format:

```
[call, put] = blsprice(price, strike, rate, time, volatility)
```

For example, for the data: stock price S0 = 42, strike price K = 40, interest rate r = 10%, maturity time T = 6 months = 0.5 years, and a volatility σ = 20%, we have:

```
[call, put] = blsprice (42, 40, 0.1, 0.5, 0.2)
```

which gives the results for the call and put options as:

```
call = 4.7594              put = 0.8086
```

The layout for the GUI to carry out this computation is shown in Figure 8.11. To this layout we change the strings for each **Static Text**, each **Edit Text**, and the **Push Buttons** as shown in Figure 8.12. Then, we change the tags for the **Edit Text** boxes with the first word in the name of the **Static Text** box next to each of them. That is, the **Edit Text** next to **Stock price** has the tag equal to **Stock**, and so on. For the **Static Text** boxes we also give the tag name in the same way. Then the top **Static Text** box next to **call** option we make the tag equal to **call** and the other one has the tag equal to **put**. For the **Push Buttons** we give the tags **Calculate** and **Close**.

We save the GUI as **BlackScholes**. Now we edit the callback for the button **Close** as in the previous example. In the callback for this **Push Button** we add:

```
close(gcbf)
```

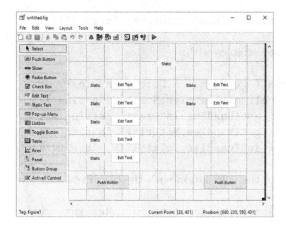

FIGURE 8.11: GUI layout.

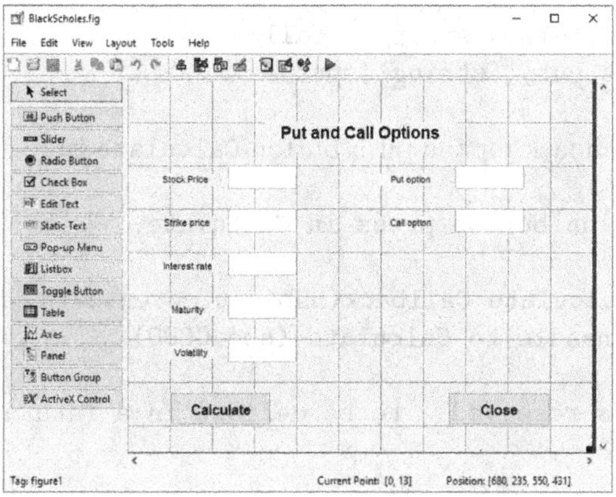

FIGURE 8.12: Final GUI layout.

The next step is to execute the Black-Scholes equation in the callback for the button **Calculate**. First we read the data from the **Edit Text** boxes and then execute the instruction **blsprice**. Finally, we write the results to the empty **Static Text** boxes. The callback for the push button **Calculate** is now described:

1. First we read in the variables for the Black-Scholes instruction **blsprice**. We do this with the instructions **eval** and **get**. As we explained above, **get** reads the string from the **Edit Text** and **eval** converts the string to a numerical value assigned to the variable **stock**. To read the variable from the **Edit Text** box Stock we use then:

```
stock = eval( get(handles.Stock, 'string'));
```

We read the five variables with:

```
stock = eval( get(handles.Stock, 'string'));
strike = eval( get(handles.Strike, 'string'));
int = eval( get(handles.Interest, 'string'));
mat = eval( get(handles.Maturity, 'string'));
vol = eval( get(handles.Volatility, 'string'));
```

2. Now, we make the calculation with the `blsprice` solution with:

```
[call, put] = blsprice(stock, strike, int, mat, vol)
```

3. We write the results to the empty `Static` text boxes with

```
set(handles.call, 'string', call)
set(handles.put, 'string', put)
```

The complete callback for the push button `Calculate` is:

```
% Executes on button press in Calculate.
%
function Calculate Callback(hObject, eventdata, handles)
% hObject handle to Calculate (see GCBO)
%
% eventdata reserved - to be defined in a future version
% of MATLAB
%
% handles structure with handles and user data (see GUIDATA)
stock = eval( get(handles.Stock, 'string'));
strike = eval( get(handles.Strike, 'string'));
int = eval( get(handles.Interest, 'string'));
mat = eval( get(handles.Maturity, 'string'));
vol = eval( get(handles.Volatility, 'string'));
[call, put] = blsprice(stock, strike, int, mat, vol);
set(handles.call, 'string', call);
set(handles.put, 'string', put);
```

Now we execute the GUI by clicking on the `Run` icon. Then we enter the values required and the results are shown in Figure 8.13.

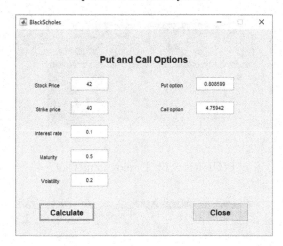

FIGURE 8.13: Final GUI with data.

8.3 Deployment of MATLAB Graphical User Interfaces

MATLAB allows deployment of MATLAB files so they can be distributed and used by other users without having a MATLAB license. These deployed files can be used from EXCEL, .NET, Java, or as stand-alone executable files. MATLAB has a tool called `Deployment tool` that guides us in the making of executable files.

In order to run an application outside of MATLAB, the end user needs to install the **MATLAB Compiler Runtime** also known as `MCR`. This is a set of functions that the executable generated uses to run. Thus, it is compulsory to install it. It can also be downloaded free of charge from the Mathworks web page "www.mathwoks.com".

There are three steps that have to be followed to obtain an executable file from a GUI which is composed of m-files and fig-files. These are:

1. Create a project.

2. Add the files.

3. Build the executable file and pack the project.

Now we describe each one of the steps:

1. We start the `Deployment Tool` by selecting the `APPS` tab in the main MATLAB window. There we select the `APPLICATION DEPLOYMENT` set and click on the `Application Compiler` icon. We can also type `deploytool` at

FIGURE 8.14: Deployment tool.

FIGURE 8.15: Loading the files to be deployed.

the Command Window. This opens the deployment menu. Then, we select the icon for New Project and which type of executable we wish to make.

2. We fill out the information requested such as application name (BlackScholesSolution) and additional details about it as shown in Figure 8.14. We save the project as BlackScholesSolution.prj. Now we add the files we need in our project by clicking on the plus sign next to the message Add main file. We work with the interface developed in Example 8.2. We start the process by adding the m-file BlackScholes.m. The file BlackScholes.fig and every other file needed are automatically loaded. We also check the radio button to include the MCR in the package if the user does not have it already (see Figure 8.15).

3. Once we have the required m-files, we proceed to build the project. We do this by clicking on the √mark located in the top right corner of the deployment window. When the process starts, the window shown in Figure 8.16 is opened. It is related to the deploying steps and it goes from Creating the binaries, to Packing, and Archiving.

4. When the process is finished, we have a set of folders with the required executable file and the MCR packed and ready for installation. The parent folder

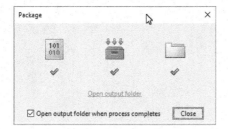

FIGURE 8.16: Creating the executable file to be deployed.

is the `BlackScholesSolution` folder and it contains three folders as follows:

- `for_redistribution`.

  It contains the file `MyAppInstaller_mcr` which installs the `MATLAB Compiler Runtime` (MCR).

- `for_redistribution_files_only`.

  It contains the executable file `BlackScholesSolution.exe`, an icon, a picture that is displayed before opening the application, and a readme.txt document with installation instructions.

- `for_testing`.

  It contains the executable file `BlackScholesSolution.exe`, a readme.txt file with installation information, the image shown when opening the application, and two other files related to the deployment process.

The folders we need to distribute are the first two: `for_redistribution` and `for_redistribution_files_only`.

5. Now we execute the file `BlackScholesSolution.exe` to open the window requesting the input data. A run produces the same window corresponding to the application shown in Figure 8.13. As we see, it is very easy to deploy MATLAB programs to use in a computer that does not have a MATLAB license.

8.4   Concluding Remarks

In this chapter we have presented the techniques to create graphical user interfaces, known as GUIs, that eases the process to execute a MATLAB program. We have presented three examples which are representative of typical GUIs. We have also presented the techniques to deploy GUIs and to create executable files that can be run in a platform without a MATLAB license.

Chapter 9

Simulink

Simulink is a package that is useful to model, simulate, and analyze systems, either in the continuous-time or discrete-time domain. They may also be either linear or non-linear, and the systems modeled in Simulink may be a combination of them. Simulink has a graphical interface that makes it very easy to build models and then simulate them. It also has a set of components grouped in libraries in different topics of engineering. Many of the libraries can be used in different topics, such as physics, finance, mechanical engineering, hydraulics, etc. Many of these libraries can be found in toolboxes which are available separately.

9.1 The Simulink Environment

To start working with Simulink we need to be working first with MATLAB. In the MATLAB main window we can either write `simulink` in the `Command Window` or click on the Simulink icon in the MATLAB main toolbar. The Simulink icon is shown in Figure 9.1. This action opens the Simulink menu window where we select the action we wish to follow. In our case we need to create a new model, as shown in Figure 9.2. After doing this, we obtain the model main window shown in Figure 9.3. By selecting from the menu window `View → Library Browser`, the Simulink library browser is opened showing

FIGURE 9.1: Simulink Icon.

all the model blocks and toolboxes available for modelling and simulation, as shown in Figure 9.4. In the context of Simulink, the toolboxes are also called blocksets. The libraries, toolboxes, and blocksets available for modeling and simulation depend upon which ones have been bought and the user needs. In this chapter our approach is on trying to get acquainted with Simulink details and we leave the details of particular libraries for the interested reader to consult the references.

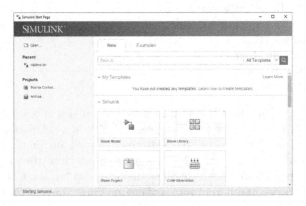

FIGURE 9.2: Simulink menu window.

9.1.1 A Basic Example

To create a new model for a system, in the `File` menu we select `New` (alternatively we can click on the New icon). This opens a model window. Then, we open the library we need and drag the required blocks to our model window. The blocks are connected together with lines. Finally, we add source signals to excite the system model and we also add sinks to observe the outputs. With a simple low pass transfer function we show the procedure.

Example 9.1 Simple model for a low pass filter
Let us consider the second-order transfer function

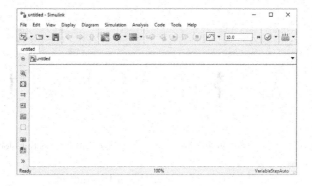

FIGURE 9.3: Simulink model window.

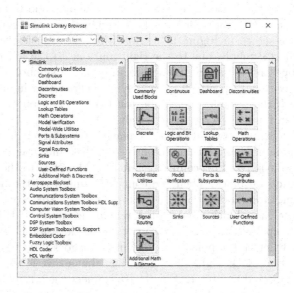

FIGURE 9.4: Simulink libraries window.

$$N(s) = \frac{1}{s^2 + s + 1}$$

This transfer function appears frequently in engineering and mathematics. For example, it can be found in electronics where it represents a control system transfer function, in physics it refers to a damped string and to a swinging pendulum, in finance a cash flow, to name a few.

This transfer function is excited by a step input and we wish to observe the output. We open a new model window and from the `Continuous library` we drag to it a `Transfer function` block, from the `Sources library` we drag a

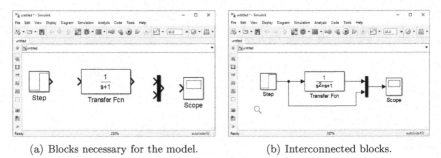

(a) Blocks necessary for the model.   (b) Interconnected blocks.

FIGURE 9.5: Creation of the model.

Step signal, from the Sinks library we drag a Scope, and from the Signal Routing library we drag a Mux block. The model window with the blocks just dragged is shown in Figure 9.5a. Note that the transfer function block has two small > symbols, one to the right and one to the left of the block, while the Scope and the Step input have only one > symbol in the left side and in the right side, respectively. The > symbols are either inputs or outputs to the blocks. If the > symbol points outwards it is an output and if it points inwards it is an input. Now, we need to connect the blocks.

The way to connect two blocks is by placing the pointer in the output of a block, the pointer changes to a crosshair, left-click and drag it to the input of the desired block. We first connect a line from the Step to the input of the Transfer Function block. Then, we connect a line from the output of the Transfer Function block to one of the inputs of the Mux block. To connect the Mux to the Step block connect the remaining input to the connection between the Step and the Transfer Function block. Another way to do this connection is to place the pointer on the line that goes from the Step to the Transfer Function block, right-click and when the crosshair appears we drag it to the unconnected Mux input. Finally, we connect the output from the Mux to the input of the Scope. The model now is shown in Figure 9.5b. This step concludes the connections among the model blocks.

Now we need to enter the correct transfer function in the Transfer Function block. We double-click on the block to open the Properties window where we change the coefficients in the denominator polynomial from [1 1] to [1 1 1] as shown in Figure 9.6. The next step is the simulation of the model. We just click on the Run icon in the model window. When the simulation ends, we click on the Scope block and the window for the Scope signals opens showing the input and output signals for the transfer function. Figure 9.7 is produced after clicking on the Autoscale icon in the Scope toolbar. We note that the simulation is up to 10 seconds and starts at t = 1 sec. The example above shows the process we have to follow to model and simulate a system. In the following sections we introduce other characteristics of Simulink. We also

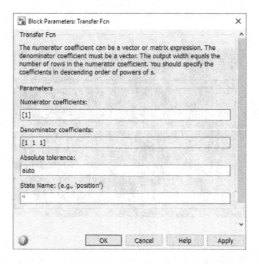

FIGURE 9.6: Properties for the `Transfer Function` block.

see that there is a toolbar on the `Scope`. The most important and useful are the `Scale y-axis limits` and the `Zoom` tools which are self-explanatory.

9.2 Continuous and Discrete Systems

A continuous model is composed of continuous blocks. These blocks are available in the `Continuous` library. Blocks from other libraries can also be used in continuous systems with the exception of the blocks in the `Discrete` and the `Logic and Bit Operations` libraries. In continuous systems the time varies continuously and in the frequency domain they use the complex frequency (Laplace transform) variable s. A discrete model uses blocks from the `Discrete` library as well as blocks from any other library with exception of blocks from the `Continuous` library. In discrete systems the time changes in steps `T` seconds known as the Sample time and it is not continuous. A system that has blocks from `Continuous` and `Discrete` libraries is called a Hybrid or Mixed-mode system. In these models the time can be either continuous or discrete. We present examples of these three kinds of systems.

Example 9.2 A continuous system
We already have a continuous model in Example 9.1. Another example is the modeling of the behavior of a pendulum. The pendulum differential equation for the case of a small swing is

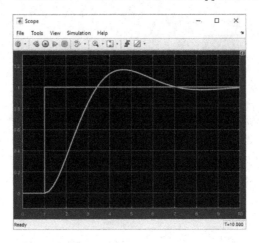

FIGURE 9.7: Input and output signals in Scope.

$$mL\ddot{\theta} + BL\dot{\theta} + W\theta = 0.04\delta(t)$$

where θ is the angular position, $\dot{\theta}$ is the angular velocity, m is the pendulum mass, $L = 0.6$ m is the length of the pendulum arm, $B = 0.08$ Kg/m/s is the damping coefficient, and $W = mg = 2$ Kg is the weight. We are assuming that $g = 9.8$ m/s^2 and $m = 0.2041$ Kg. This differential equation can be written as a set of linear differential equations. First define the state-variables $x_1(t) = \theta$ and $x_2(t) = \dot{\theta}$. Then, we can write the differential equation as a system of first-order differential equations:

$$
\begin{aligned}
x_1 &= x_2 \\
\dot{x}_2 &= -16.33x_1 - 0.392x_2 - 0.04\delta(t)
\end{aligned}
$$

This system of equations can be written in matrix form as

$$
\begin{bmatrix} \dot{x}_1 \\ \dot{x}_2 \end{bmatrix} = \begin{bmatrix} 0 & 1 \\ -16.33 & -0.392 \end{bmatrix} \begin{bmatrix} x_1 \\ x_2 \end{bmatrix} + \begin{bmatrix} 0 \\ -0.04\delta(t) \end{bmatrix}
$$

$$
\begin{bmatrix} y_1 \\ y_2 \end{bmatrix} = \begin{bmatrix} 0 & 1 \end{bmatrix} \begin{bmatrix} x_1 \\ x_2 \end{bmatrix}
$$

The initial conditions are

$$
\begin{bmatrix} x_1(0) \\ x_2(0) \end{bmatrix} = \begin{bmatrix} \pi/2 \\ 0 \end{bmatrix}
$$

This system is modeled in Simulink with the **State-Space** block from the
Continuous library. The input to this block is an impulse and the output
is seen in a **Scope**. The complete model is shown in Figure 9.8a. For the
State-Space block we enter the system matrix A and the vectors B, C, and
D together with the initial conditions as shown in Figure 9.8b. The impulse
is created with two step signals subtracting the second from the first but the
second one delayed 0.01 sec from the first one that has the step at 0 sec. After
running the model we see the response as shown in Figure 9.9a. The output
signal is not as good as we might expect. This is due to the fact that in the
simulation the step is set to **Auto** as can be seen in Figure 9.9b which is opened

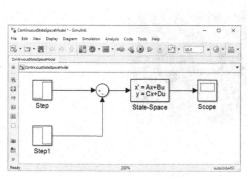 

(a) Complete continuous model.

(b) **State-Space** block parameters.

FIGURE 9.8: Continuous model.

from the **Simulation** menu with **Configuration Parameters**. Changing the
maximum step size from **Auto** to 0.01, clicking on the OK button and running
the simulation again we get the output shown in Figure 9.9c which shows
a smoother waveform. From the output we conclude that the pendulum decreases the amplitude of the oscillations and that eventually it will stop.

Example 9.3 A discrete system
A discrete system has blocks from the **Discrete** library. Discrete models appear in digital signal processing (DSP), at discrete-time signal processing such
as in finance, switched-capacitor filters in electronics, finite difference methods in engineering, to mention a few. The variable in discrete models is the
z-transform variable z and the time is in discrete steps. As an example, let
us consider the model shown in Figure 9.10a. It is a digital filter and the basic block is the unit delay block represented by $1/z$. This model realizes the
transfer function

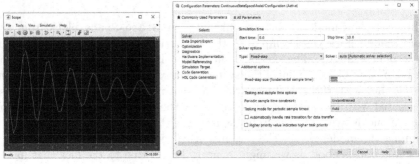

(a) Poor sampling.   (b) Changing the time step.

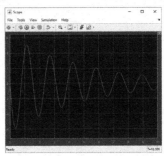

(c) Improved output signal.

FIGURE 9.9: Output signal.

$$H(z) = \frac{z^2 - 1}{z^2 - 0.75z - 0.794}$$

The response is shown in Figure 9.10b and we see that it is a discrete time waveform. The input signal is a unity frequency signal with an amplitude of 1 and a sample time of 0.5 s.

Example 9.4 A mixed-mode model

For an example of a mixed-mode system let us consider the cascade of a continuous transfer function block and of a discrete-time one. The system excitation is a step signal that starts at 0 sec. The model is shown in Figure 9.11a and the response in Figure 9.11b. We readily see that the response is a discrete-time signal.

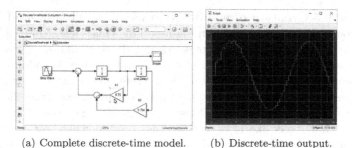

(a) Complete discrete-time model. (b) Discrete-time output.

FIGURE 9.10: Discrete-time model.

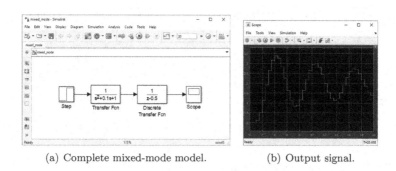

(a) Complete mixed-mode model. (b) Output signal.

FIGURE 9.11: Mixed-mode model.

9.3 Subsystems

Similar to the case of functions in MATLAB and subroutines in other programming languages, portions of a model can be grouped in a model for reuse and customization. This is especially useful when the model has grown significatively. A subsystem is a hierarchical grouping of blocks encapsulated by a single subsystem. A subsystem can have many blocks that could make it hard to understand the complete model. A subsystem can be reusable, thus if the same subsystem function is needed again, we can repeat the subsystem within the model.

There are two ways to create a subsystem. In the first method to create a subsystem, we only need to select a group of blocks and then from the `Diagram` menu select `Subsystem & Model Reference` and then `Create Subsystem from Selection` (or Control-G). In the second method, we start with an empty `Subsystem` block available in the library `Ports & Subsystems` and we place the blocks and lines needed for the subsystem. We show these procedures with examples.

Example 9.5 Creation of a subsystem from an existing model

Let us consider the model shown in Figure 9.12a. This is a state-variable filter. The first step to create a subsystem is to select the parts of the system that we wish to have in the subsystem. For our example, these blocks and lines for the subsystem are shown in Figure 9.12b. Next, from the menu we select `Diagram` → `Subsystem & Model Reference` → `Create Subsystem from Selection` as can be seen in Figure 9.12c. This action creates the subsystem and the model window looks like it is seen in Figure 9.12d where we see that the components of the model selected before are now inside the subsystem. Now we can run the model by clicking on the `Run` icon.

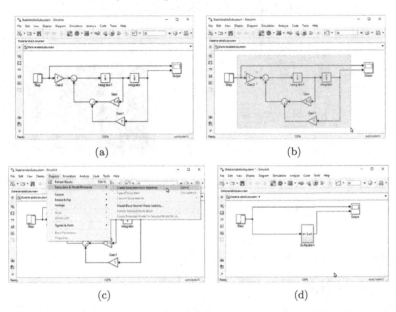

FIGURE 9.12: Subsystem from an existing model, a) Original model, b) Selection of the blocks and lines for the subsystem, c) Path to create the subsystem, d) Main model window.

Example 9.6 Creating a subsystem using the Subsystem block

To create a subsystem in this way, in a model window we drag a `Subsystem` block from the `Ports & Subsystems` library, as shown in Figure 9.13a. We double-click on the block and see that there are only input and output ports and a line connecting them as we see in Figure 9.13b. We edit the subsystem model as needed. In this example we add the state-variable filter as shown in Figure 9.13c. We close the subsystem model window and return to the main model window where we add a `Pulse` generator from the `Sources` library and a `Scope` from the `Sinks` library (see Figure 9.13d). Finally, we run the simulation by clicking on the `Run` icon or by Control+T.

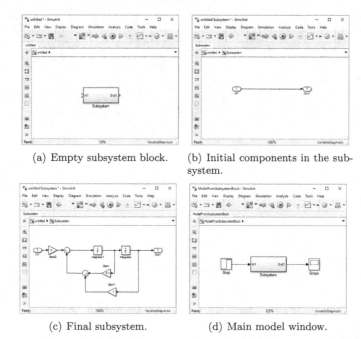

(a) Empty subsystem block.

(b) Initial components in the subsystem.

(c) Final subsystem.

(d) Main model window.

FIGURE 9.13: Subsystem from a subsystem block.

9.3.1 Masking Subsystems

Masking enables us to use a subsystem as an independent block, thus extending the concept of abstraction. A masked block can have a custom icon and a dialog box of its own where we can enter parameters as if it were a Simulink block. Each and every block available in the Simulink libraries is masked. The procedure to mask a subsystem consists in the following steps:

1. We first build a subsystem.

2. In the main model window, we select the subsystem block.

3. Select the subsystem block and select Diagram → Mask → Create Mask.... The Mask Editor: Subsystem window is open. The tabs in this window are shown in Figure 9.14.

4. In the Mask Editor Subsystem window, we work in the Parameters & Dialog tab and in the Documentation tab.

Let us consider again the state-variable filter subsystem shown in Figure 9.13c. Now we change the values of the Gain blocks to A0 and A1. This can be

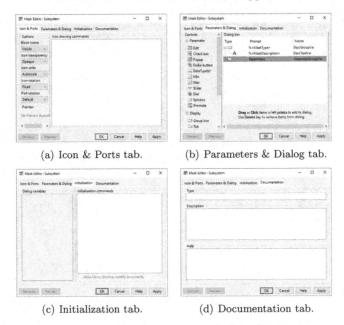

(a) Icon & Ports tab.   (b) Parameters & Dialog tab.

(c) Initialization tab.   (d) Documentation tab.

FIGURE 9.14: Mask Editor pane.

done by double-clicking on each of the `Gain` blocks and changing the `Gain` parameters from their values to `A1` and `A0` as shown in Figure 9.15a for the `Gain` block `A1`. The final subsystem is shown in Figure 9.15b. Now, the `Gain` parameters of the model are the values of the `Gain` blocks whose values are given by the coefficients of the transfer function. We now mask this subsystem in such a way that the values of the gain blocks are given in a dialog box. In the main model window of Figure 9.13d we select the subsystem block and then `Diagram` $\rightarrow$ `Mask` $\rightarrow$ `Create Mask....` This opens the dialog window for the Mask Editor for a subsystem. The mask editor has four tabs: `Icon`,

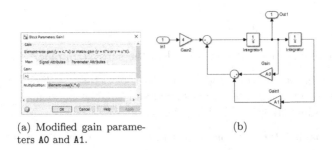

(a) Modified gain parameters `A0` and `A1`.

(b)

FIGURE 9.15: Editing the `Gain` blocks to `A0` and `A1`.

Parameters, Initialization, and Documentation, as shown in Figure 9.14. We now describe each of these tabs.

9.3.2 Icon & Ports Tab

The Icon & Ports dialog window allows us to create an icon for the subsystem block. As can be seen in Figure 9.14a, there is a window where we enter the commands that draw the icon on the subsystem block. Some of the options available are shown in Table 9.1.

TABLE 9.1: Options for the Icon and Ports tab.

Option	Description
Plot a curve	Use MATLAB plotting instructions.
Display an image	The image must be in the model's folder.
Name the I/O ports	Use the instruction port_label('input' or 'output', port number,'label')
Transparency	Set the Transparency pulldown menu in the Icon options column.
Rotation	Set the Rotation pulldown menu in the Icon options column.
Change of units	Set the Units pulldown menu in the Icon options column.
Change the color	Use the color code from MATLAB. Place before an instruction.
Display a transfer function	Use poly(numerator, denominator).
Plot poles and zeros	Use droots(zeros vector, poles vector, constant)
Print a text	Use text('text to write')
Print a formatted text	Use fprintf as in MATLAB.

As examples the reader could try the following commands:

- port_label('input', 1,'Filter in')
- color('blue')
- text(0.5, 0.5, 'State-Variable')
- plot([1,0], [0.3, 0.8])

9.3.3 Parameters & Dialog Window

The Parameters & Dialog pane is shown in Figure 9.14b. In this pane we enter the parameters for the subsystem which in our example are the coefficients A0 and A1. To do this we click on the Edit button at the Controls

FIGURE 9.16: Parameters & Dialog tab.

FIGURE 9.17: Prompt for parameters A0 and A1.

column to add a new parameter to the now empty list. At the **Property editor** at the right we change the **Name** to **A0**, **Value** to 1, and at **Prompt** we enter **Coefficient A0**. We repeat the procedure for **A1**. We add **A1** at **Name**, at **Value** we add 1, and at the **Prompt** box we enter **Coefficient A1**. The final window is shown in Figure 9.16. These two parameters will appear at the dialog window for the block when we double-click on it once the masking procedure has been finished. After we press the button **Apply** we double-click on the subsystem block and the dialog window of Figure 9.17 will prompt for the values of the coefficients **A0** and **A1**.

9.3.4 Initialization Tab

In this window we can enter MATLAB instructions that allow an initialization of the subsystem block. The block is initialized when:

> The model is loaded.
> The simulation is started or the block is updated.
> The block is rotated.
> The block's icon is redrawn.

We can additionally enter MATLAB commands in the **Initialization Commands** box to initialize the masked subsystem.

FIGURE 9.18: Description and Help.

9.3.5 Documentation Tab

The `Documentation` tab window is shown in Figure 9.14d. This tab allows us to define the model parameters, the description, and the help text for the masked subsystem block. For example, for our subsystem we can enter the following text in the `Mask type` dialog box:

`State-variable filter model.`

In the `Description` box we can enter:

`This is a realization in state-variable form of a second order transfer function.`

Finally in the `Mask help` we can enter:

`Enter the coefficients of the transfer function denominator. A0 is the independent coefficient and A1 is the linear term coefficient.`

After clicking on the `Apply` button, we double-click on the masked subsystem block and obtain the `Block Parameters` dialog window for the block shown in Figure 9.18.

9.4 Examples

We present examples of different systems that can be modeled in Simulink. The first example is a digital circuit, this example shows another Simulink strength that has been overlooked for some time, namely, the simulation of digital circuits. The second example is an application in finance which can be modeled as a discrete-time model.

Example 9.7 Modeling of a digital sequential circuit

Let us consider the counter with unused states defined by the state diagram of Figure 9.19a. A Simulink model for this counter is shown in Figure 9.19. We are using the JK flip-flops that are available in the library `Simulink Extras`. The JK flip-flops can be initialized to an initial state of 1 or 0. In our circuit we initialize the flip-flops to 1. This can be done by double-clicking on the flip-flops and entering a unity value for the initial. The JK flip-flops are negative edge triggered. The digital clock is set to a `Period` = 0.5. We now run the simulation and see the output signals at the `Scope` and shown in Figure 9.20. There we can see that the initial state for bits A, B, and C is 111. (The least significant bit is A.) When the first negative edge clock happens the counter changes state to 000. The reader can observe that after each negative edge the output changes to another state according to the state diagram of Figure 9.19.

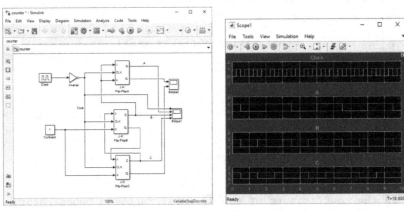

(a) Simulink model of the digital counter.   (b) Digital counter output signals.

FIGURE 9.19: Digital counter with unused states.

Example 9.8 Savings for retirement

A 25-year-old man is planning to retire when he reaches the age of 65 years. He begins a savings plan starting with an initial deposit of $10000. He wishes to know how much he is going to have at the end of the his productive life if he saves $200 each month at an annual interest rate of 6%. The data for this problem is: Initial condition = 10000, monthly interest rate= 0.5, months in plan= 40×12= 480. The problem can be modeled by the difference equation

$$\text{sum}(nT) = \text{sum}[(n-1)T] * \text{interest\_rate} + \text{monthly\_deposit}$$

A Simulink model that realizes this equation is shown in Figure 9.20a. Here, we have a `constant` block as input with a value of 200 which is the monthly deposit. In addition, we have a time delay which represents the month that passes between deposits. The feedback amplifier represents the increase in value of the deposit due to the monthly interest. This is added to the monthly deposit to have the new sum. Finally, we are displaying the sum in the `Scope`, in a `Display` block and in a `To Workspace` block (both blocks are available in the `Sinks` library) to send the numerical data to MATLAB. The initial conditions in the `Delay` block is set to 10000 to account for the initial deposit. The `Gain` block value is 1.005 which is the original amount plus the monthly interest of 0.5 %. The final model is shown in Figure 9.20a. The simulation is run for 480 seconds to account for the 480 months in the plan. We now run the simulation and observe that the display block shows that the final amount saved is $507,872.68. Note that if the money were not invested, the savings at the end of the plan would have only been $106,000. If we now go to MATLAB we see that in the `Workspace` window we have two variables, `simout` and `tout`. `tout` is a vector of length 480 which is the number of months. `simout` is an array which has three elements: `time`, `signals`, and `blockName`. Of these elements, `signals` has three components, one of them being `values`. If we use the icon plot on `values` we get an identical plot as compared to the `Scope` output shown in Figure 9.20b.

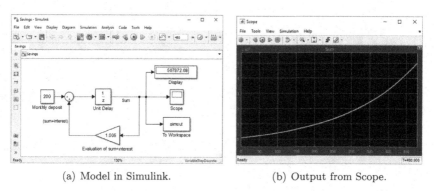

(a) Model in Simulink.   (b) Output from Scope.

FIGURE 9.20: Model for retirement savings.

9.5 Concluding Remarks

This chapter covers a very useful and important application of MATLAB. This application is called Simulink and it is block-design oriented. The examples presented show that it is very easy to use. In the same way as MATLAB, it can be enriched with toolboxes that are especially designed for certain areas of science, and engineering. An important topic presented is the creation of subsystems and the masking of these subsystems. Finally, some examples show some of the capabilities of Simulink.

9.6   References

[1] J.B. Dabney and T.L. Hartman, Mastering Simulink, Prentice Hall, Inc., Upper Saddle River, NJ, 2004.

[2] S. Karris, Introduction to Simulink with Engineering Applications, Third Edition, Orchard Publications, 2011.

[3] H. Klee and R. Allen, Simulation of Dynamic Systems with MATLAB and Simulink, CRC Press, Boca Raton, FL, 2017.

[4] M. Nuruzzaman, Modeling and Simulation in SIMULINK for Engineers and Scientists, AuthorHouse, Bloomington, IN, 2005.

[5] Simulink Users Guide, The MathWorks, Inc., Natick, MA, 2018.

Chapter 10

MATLAB Applications to Engineering

10.1 Introduction

In this chapter we present some applications to branches in engineering. Electronic, civil, mechanical, chemical and food engineering applications are covered. Of course, we do not pretend to be exhaustive because that would be impossible to do. However, we present some representative examples in those areas. The examples presented include applications of some of the toolboxes available from The Mathworks, Inc. while other examples have scripts written specifically for them. Among the toolboxes used we have the Optimization, Symbolic Math, Control, and Signal Processing toolboxes.

10.2 Applications to Signals and Systems

Frequently, engineering systems are modeled either by a differential equation or by a transfer function. Any one of these descriptions provide a great deal of information about the system. For example, we can obtain information about the stability of the system, its step or impulse response, and its frequency response. This section uses the Signal Processing toolbox.

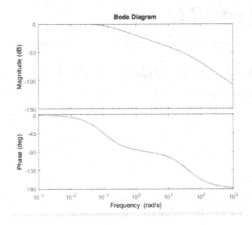

FIGURE 10.1: Bode plots.

Example 10.1 Second order transfer function behavior.
Bode plots are plots of magnitude in dB and phase in degrees of the transfer function $N(s)$. Let us consider the second order transfer function

$$N(s) = \frac{\omega_n^2}{s^2 + 2\omega_n\xi s + \omega_n^2} \tag{10.1}$$

where ω_n and ξ are the natural frequency and damping coefficient, respectively. This transfer function has applications in control and analog filtering. We wish to know the response of $N(s)$ when $\omega_n = 2$ and $\xi = 0.1$. For these values, $N(s)$ is then,

$$N(s) = \frac{4}{s^2 + 40s + 4}$$

We can obtain a plot of its magnitude and phase with the instruction **bode**. This instruction produces plots of magnitude in dB and phase in degrees. The instruction **bode** requires to give the numerator and denominator polynomials. For our function we have

$$\texttt{num = [4]}$$

$$\texttt{den = [1 40 4]}$$

Then, the instruction **bode** is

$$\texttt{bode(num, den)}$$

This will produce the Bode plots shown in Figure 10.1. Note that the function $N(\texttt{s})$ corresponds to a low pass filter. Also note that the phase is negative. It has a minimum value of $-180°$. This is so because the transfer function has

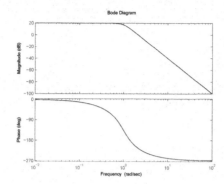

FIGURE 10.2: Bode plots for a third order function.

two poles and two zeros at infinity, and each pole contributes a maximum of $-90°$ to the total phase. Now, let us consider the plot for the function

$$N(s) = \frac{10}{s^3 + 2s^2 + 2s + 1}$$

The Bode plots for this function are shown in Figure 10.2. In this case the phase plot minimum value is $-270°$ because the transfer function has three poles.

If we now wish to see the behavior of the second order function for different values of ξ we can use the following script that uses a `for` instruction to increment ξ values,

```
% This is file Example10_1.m
clc
clear
close all
num = [1];
zeta = [ 0.1 0.2 0.3 0.4 ];
for k = 1:4
      den = [1 2*zeta( k ) 1];
      bode(num, den);
      hold on
end
hold off
```

The result is shown in Figure 10.3. There we see that the effect of changing the parameter ξ affects the peak in the magnitude and the phase around the cutoff frequency.

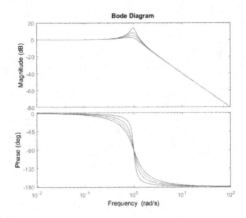

FIGURE 10.3: Effect of changing ξ in a second order transfer function.

Example 10.2 Time domain response of a second order function.
The unit-impulse and the unit-step response of the second order transfer function can be obtained with the instructions `impulse` and `step` which have the format

$$\text{step(numerator, denominator)}$$

For the second order transfer function

$$N(s) = \frac{16}{s^2 + 2s + 16}$$

we have then that the impulse and step responses can be obtained with

```
This is file Example10_2A.m
num = [16];
den = [1 2 16];
impulse(num, den)
figure
step(num, den, ':')
legend('Impulse response', 'step response')
```

which produces the plots shown in Figure 10.4. We now wish to know the unit-step response of the second order transfer function when ξ changes. For example, for the values of $\xi = 0.1, 0.2, 0.3,$ and 0.4, we can plot the step response with the following script:

```
This is file Example10_2B.m
num = [1];
```

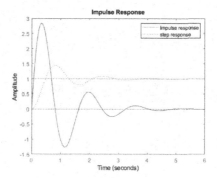

FIGURE 10.4: Unit-impulse and unit-step responses.

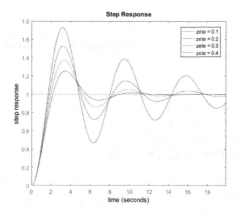

FIGURE 10.5: Step response for different values of ξ.

```
for zeta = 0.1:0.1:0.4
   den = [1 2*zeta 1];
   t = 0:0.1:19.9;
   step(num, den, t)
   hold on
end
xlabel('time')
ylabel('step response')
legend('zeta = 0.1','zeta = 0.2','zeta = 0.3','zeta = 0.4')
```

We can readily see in Figure 10.5 that the overshoot is larger for smaller values of ξ. The same information available in Figure 10.5 for the step response can be plotted in a three-dimensional plot with:

```
This is file Example10_2C.m
```

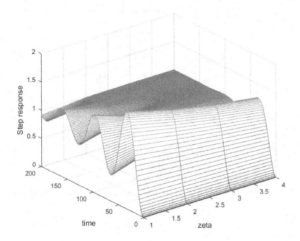

FIGURE 10.6: Mesh plot of step response for different values of ξ.

```
k = 1;
for zeta = 0.1:0.1:0.4
    den = [1 2*zeta 1];
    t = 0:0.1:19.9;
    y(:, k) = step(num, den, t);
    k = k + 1;
end
mesh(y)
xlabel('zeta')
ylabel('time')
zlabel('Step response')
```

The plot shown in Figure 10.6 gives an idea of how the step response changes with the parameter ξ.

Example 10.3 Laplace Transforms.
The Laplace transform of a piecewise continuous function $f(t)$ for $t \geq 0$ is defined by

$$\mathcal{L}\{f(t)\} = F(s) = \int_0^\infty f(t)e^{-st}dt \qquad (10.2)$$

where s is the complex frequency variable $s = \sigma + j\omega$. On the other hand, the inverse transform is given by

$$f(t) = \frac{1}{2\pi j} \int_{c-j\infty}^{c+j\infty} F(s)e^{st}ds \qquad (10.3)$$

where c > 0. MATLAB allows us to obtain the direct and inverse Laplace transforms of $f(t)$ with the instructions `laplace` and `ilaplace`, respectively. For example, for the functions

$$f_1(t) = t^3 e^{-3t} \qquad f_2(t) = \cosh(2t)e^{-4t}$$

the Laplace transforms are given by

```
This is file Example10_3.m
syms t a w
f1 = t^3*exp(-3*t);
f2 = cosh(2*t);
F1 = laplace( f1 )
F2 = laplace( f2 )
```

to obtain the results

```
F1 =
     6/(s + 3)^4
F2 =
     s/(s^2 - 4)
```

For the inverse Laplace transforms we have that

```
>> g1 = ilaplace(F1), g2 = ilaplace(F2)

g1 =
     t^3*exp(-3*t)

>> g2 =
     exp(-2*t)/2 + exp(2*t)/2
```

that can be simplified to

$$\cosh(2t)$$

Thus, we obtain the original functions.

A property for Laplace transforms says that the derivative of a function has the Laplace transform

$$\mathcal{L}\left\{\frac{df(t)}{dt}\right\} = sF(s) - f(0) \qquad (10.4)$$

If we have a function with vanishing initial conditions, then we can use the instruction `laplace` to find the transform of a derivative. For example, for the function

$$f(t) = t\,\sin(t)$$

its Laplace transform is

```
>> F = laplace( f )

F =

2*s/(s^2 + 1)^2
```

Then, the Laplace transform of its derivative is

```
>> laplace(diff(f))

ans =

    2*s^2/(s^2+1)^2
```

We can clearly see that the Laplace transform of $f(t)$ is multiplied by s when we first take the derivative of $f(t)$ and then the Laplace transform.

Example 10.4 Fourier transforms
In a similar way to the evaluation of Laplace transforms, we can obtain the Fourier transform of a function as well as its inverse Fourier transform. The Fourier transform of a function $y(t)$ is defined by

$$\mathcal{F}\{y(t)\} = Y(j\omega) = \int_{-\infty}^{\infty} y(t)e^{-j\omega t}dt \qquad (10.5)$$

and the inverse Fourier transform is

$$\mathcal{F}^{-1}\{F(j\omega)\} = f(t) = \frac{1}{2\pi}\int_{-\infty}^{\infty} F(j\omega)e^{j\omega t}d\omega \qquad (10.6)$$

The MATLAB instructions to get the Fourier and the inverse Fourier transforms are **fourier** and **ifourier**, respectively. For example, for the function

$$f(t) = e^{-3|t|}$$

its Fourier transform is then obtained with

```
>> syms t
>> f = exp(-3*abs(t)) ;
>> fourier(f)

ans =
    6/(w^2 + 9)
```

As another example, for the function $F(w) = \pi(w - 3)$, the inverse Fourier transform is

```
>> F = pi*(w - 3);
>> f = ifourier(F)

f =
   -pi*(-6*pi∧2*dirac(x) + pi∧2*dirac(1, x)*2*i)/(2*pi)
```

where the dirac function is the impulse function at x = +1 for `dirac(1, x)` and at x = 0 for `dirac(x)`.

Example 10.5 z-transform
The z-transform is defined for discrete signal sequences $x(n)$ and it is given by

$$\mathcal{Z}\{x(n)\} = X(z) = \sum_{n=-\infty}^{\infty} x(n)z^{-n} \tag{10.7}$$

The inverse z-transform is given

$$x(n) = \mathcal{Z}^{-1}\{X(z)\} = \frac{1}{2\pi j} \oint_C X(z)z^{n-1}dz \tag{10.8}$$

They can be evaluated with the instructions **ztrans** and **iztrans**, respectively. As an example let us consider the sequence $x(n) = e^{-kn}$. Its z-transform is

```
>> syms k n
>> x = exp(-k*n);
>> >> X = ztrans(x)

X =
   z/(z - exp(-k))
```

For an example in the evaluation of an inverse z-transform, let us consider the z-transform given by

$$X(z) = e^{\frac{1}{z}}$$

The inverse z-transform is obtained with

```
>> syms z
>> X = exp(1/z);
>> iztrans(X)

ans =
```

1/n!

That is, the original sequence is

$$x(n) = \frac{1}{n!}$$

10.3   Applications in Digital Signal Processing

The area of digital signal processing has also benefited from the use of the Signal Processing toolbox. We show some examples for discrete signals. We start with a convolution, then continue with the discrete Fourier transform and how we can perform convolution with it. Finally we design some digital filters with this toolbox.

Example 10.6 Linear convolution.
The linear convolution of two finite-duration (also called finite length sequences) discrete signals $x_1(n)$, and $x_2(n)$ is given by

$$y(n) = \sum_{k=-\infty}^{\infty} x_1(k)x_2(n-k) \qquad (10.9)$$

MATLAB performs a convolution with the instruction conv(x1, x2). The values of the elements of the sequences must be in a polynomial format. Thus, for the sequences

x1 = [1 2 3 4 5 4 3 2 1]   and   x2 = [1 2 -1]

we have that the convolution is a new sequence $y(n)$ of length $n_1 + n_2 - 1$ and given by

```
>> x1 = [1 2 3 4 5 4 3 2 1]; x2 = [1 2 -1];
>> y = conv(x1, x2)
 y =
   1 4 6 8 10 10 6 4 2 0 -1
```

We see that the result has a length 11 because x_1 and x_2 have lengths 9 and 3, respectively. We can plot the convolution sequence with

```
>> stem (y)
>> xlabel ('n')
>> ylabel('y(n)')
>> axis([0 12 -3 12])
```

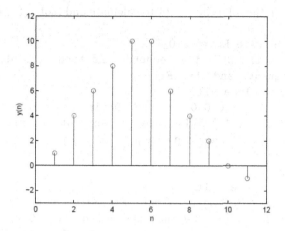

FIGURE 10.7: Sequence x_3 resulting from the convolution of x_1 and x_2.

The convolution sequence is plotted in Figure 10.7.

Example 10.7 Discrete Fourier transform.
The discrete Fourier transform (DFT) of a sequence of length N is given by

$$X(k) = \sum_{n=0}^{N-1} x(n)e^{-j2\pi kn/N} \qquad k = 0, 1, ..., N-1 \qquad (10.10)$$

The inverse discrete Fourier transform (IDFT) is given by

$$x(n) = \frac{1}{N} \sum_{k=0}^{N-1} X(k)e^{j2\pi kn/N} \qquad n = 0, 1, ..., N-1 \qquad (10.11)$$

Actually, MATLAB computes the DFT using the algorithm for the Fast Fourier Transform FFT. This algorithm is an efficient method to compute the DFT of a sequence. It optimizes the number of multiplies and additions used in the computation of the DFT. The instructions `fft` and `ifft` are used to perform the DFT and the IDFT in MATLAB. For example, for the finite length sequence

$$x_1(n) = \begin{cases} 1 & 0 \leq n < 5 \\ 0 & 5 \leq n \leq 9 \end{cases}$$

We have written the following script to calculate and plot $x_1(n)$ and the DFT:

```
%  This is file Example10_7.m
% Computes the DFT of a sequence and then it plots
% the sequence and the DFT.
clear, clc, close all
x1 = [ 1 1 1 1 1 0 0 0 0 0 ]; % Defines the sequence.
subplot ( 2, 1, 1 ) % Defines the subplot for x1.
stem(x1) % Plots the sequence x1.
axis([0 10 -1 2])
xlabel('n')
ylabel('Sequence x_1(n)')
X1 = fft (x1);% Evaluates the DFT.
k = [ 0 : 1 : 9 ]; % Defines the vector from 0 to 9.
subplot (2, 1, 2) % Defines the subplot for the DFT.
stem ( k, X1 ) % Plots the DFT of x1.
axis([0 10 -1 6])
xlabel('n')
ylabel('DFT of x_1(n)')
stem ( k, X1 ) % Plots the DFT of x1.
```

The circular convolution of two sequences is obtained by multiplying the DFT of the sequences. Thus for the sequences $x_1(n)$ and $x_2(n)$ of Example 10.6 we have

```
>> x1 = [1 2 3 4 5 4 3 2 1];
>> x2 = [1 2 -1];
>> F1 = fft(x1)
 F1 =
 Columns 1 through 3
 25.0000 -7.7909 - 2.8356i 0.2169 + 0.1820i

 Columns 4 through 6
 -0.5000 - 0.8660i 0.0740 + 0.4195i 0.0740 - 0.4195i

 Columns 7 through 9
 -0.5000 + 0.8660i 0.2169 - 0.1820i -7.7909 + 2.8356i

>>  F2 = fft(x2)
F2 =
 2.0000 0.5000 - 2.5981i 0.5000 + 2.5981i
```

We see that the two DFT's have different lengths and in order to multiply them they must have the same length, thus we pad F2 with zeros in such a way that it has the same length as that of F1. The sequence F2 is now given by

```
>> F2 = 2.0000, 0.5000, -2.5981i, 0.5000+2.5981i, 0,0,0,0,0,0,0
```

which is of length 9. Now we just multiply them as

```
>> F3 = F1.*F2

  F3 =
    50.0000 -11.2627 +18.8236i -0.3644 + 0.6545i 0 0 0 0 0 0
```

We finally take the IDFT to obtain the convolution as

```
>> f3 = ifft(F3)

f3 =
Columns 1 through 3
4.2637 + 2.1642i 3.1739 + 0.7706i 3.2917 - 0.9514i

Columns 4 through 6
4.4532 - 2.1308i 6.0319 - 2.3117i 7.3691 - 1.5077i

Columns 7 through 9
7.9498 - 0.0334i 7.4609 + 1.5411i 6.0059 + 2.4591i
```

Note that the sequence f3 has length 9, as opposed to the result in Example 10.6. This is so because the circular convolution is different from the linear convolution. However, we can perform linear convolution with the DFT just by padding with zeros the two sequences adding as many zeros as the final length of the linear convolution. Thus, in the case of the same sequences we pad with zeros so now they become

$$x1 = [1\ 2\ 3\ 4\ 5\ 4\ 3\ 2\ 1\ 0\ 0];$$
$$x2 = [1\ 2\ -1\ 0\ 0\ 0\ 0\ 0\ 0\ 0];$$

Now we perform the following computations to obtain the linear convolution of $x_1(n)$ and $x_2(n)$,

```
>> X1 = fft(x1); X2 = fft(x2);
>> X1.*X2;
>> X3 = X1.*X2
```

```
X3 =
Columns 1 through 3
50.0000 -19.9268 -19.7614i 0.1898 + 0.7092i

Columns 4 through 6
0.2700 - 4.0678i -0.1922 + 0.8645i 0.1592 - 1.2008i

Columns 7 through 9
0.1592 + 1.2008i -0.1922 - 0.8645i 0.2700 + 4.0678i

Columns 10 through 11
0.1898 - 0.7092i -19.9268 +19.7614i

>> x3 = ifft(X3)

x3 =

Columns 1 through 6
1.0000 4.0000 6.0000 8.0000 10.0000 10.0000

Columns 7 through 11
6.0000 4.0000 2.0000 -0.0000 -1.0000
```

We can see that the linear convolution obtained using the DFT is the same
as the one obtained using the instruction `conv`.

10.4   Applications in Control

MATLAB finds many applications in Control theory. Thus, an especially ded-
icated toolbox has been produced and it is available as a separate product.
It is called the Control Toolbox. In this section we present some examples in
Control theory.

Example 10.8 Stability in a feedback system.
MATLAB can perform a great deal of functions from the Control toolbox. For
example, let us consider the feedback system shown in Figure 10.8. The plant
is defined by the function:

$$G(s) = \frac{1}{s^3 + 2s^2 + 2s + 1}$$

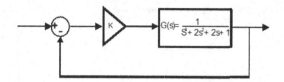

FIGURE 10.8: Block diagram of a plant with feedback.

This plant can be defined in MATLAB as

```
nump = [1];
denp = [1 2 2 1];
plant = tf(nump, denp) % It displays the transfer function.
```

and MATLAB delivers the open-loop transfer function:

```
Transfer function:
        1
  _____
  s∧3 + 2 s∧2 + 2 s + 1

Continuous-time transfer function.
```

If the system gain is K = 1, the negative feedback system is defined as

$$\text{system = feedback(plant, [1])}$$

to obtain the complete system transfer function as

```
system = feedback( plant, [ 1 ])

Transfer function:
        1
  _____
  s∧3 + 2 s∧2 + 2 s + 2

Continuous-time transfer function.
```

Note the independent coefficient in the denominator is now 2, but this is the closed-loop transfer function. To check for system stability we obtain the poles with

```
poles = pole(system)
```

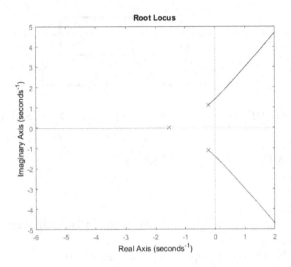

FIGURE 10.9: Root locus of a feedback system.

to get

```
poles =
-1.5437
-0.2282 + 1.1151i
-0.2282 - 1.1151i
```

We see that the system has three poles because we have a third order denominator and that the three poles are in the left-hand plane and thus, the system is stable. We can find out more on this system. For example, we can obtain its root locus. This is a plot of root loci versus the gain K. A root locus can be found with the instruction `rlocus` as in

```
rlocus(system)
```

The root locus is shown in Figure 10.9. We show two data tips and we see the conditions for stability and instability. We see that a gain of 2.36 makes the system unstable.

Example 10.9 Control of the Hubble telescope
A model for the Hubble telescope which is in orbit around the Earth, together with a positioning system is shown in Figure 10.10.

The goal is to find values for K and K2 such that the overshoot when a step input is applied is less than or equal to 5%, that the error when we apply a ramp r(t) is minimized, and that the effect of a perturbation step is reduced.

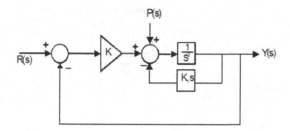

FIGURE 10.10: Model for the Hubble telescope.

The system has two inputs, $R(s)$ and $P(s)$, and then we have

$$Y(s) = \frac{KG(s)}{1 + KG(s)}R(s) + \frac{G(s)}{1 + KG(s)}P(s) \qquad (10.12)$$

The error is given by

$$E(s) = \frac{1}{1 + KG(s)}R(s) - \frac{G(s)}{1 + KG(s)}P(s) \qquad (10.13)$$

First we try to satisfy the 5% overshoot requirement. Thus, we only take into account the input-output system with a step input with value A as:

$$\frac{Y(s)}{R(s)} = \frac{KG(s)}{1 + KG(s)} = \frac{K}{s(s + K_1) + K} = \frac{K}{s^2 + K_1 s + K} \qquad (10.14)$$

From Figure 10.4, we see that for a 5% overshoot we need $\xi = 0.7$. For a ramp $r(t) = Bt$, the steady-state error is given by

$$\epsilon_{ss}^r = \lim_{s \to 0} \frac{B}{KG(s)} = \frac{B}{K/K_1}$$

And for a unit step perturbation

$$\epsilon_{ss}^r = \lim_{s \to 0} \frac{-1}{(s + K) + K} = \frac{-1}{K}$$

In order to decrease the error due to a perturbation we need to increase K. In addition, to decrease the steady-state error when the input is a ramp we need to increase K/K_1. Further we need $\xi = 0.7$. The characteristic equation is

$$s^2 + 2\xi\omega_n s + \omega_n^2 = s^2 + k_1 s + K = s^2 + 2(0.7)\omega_n s + K$$

Thus, $\omega_n = \sqrt{K}$ and $K_1 = 0.7\omega_n$, and then $K_1 = 1.4\sqrt{K}$. Therefore,

$$\frac{K}{K_1} = \frac{K}{1.4\sqrt{K}} = \frac{\sqrt{K}}{1.4} = \frac{\sqrt{K}}{2\xi}$$

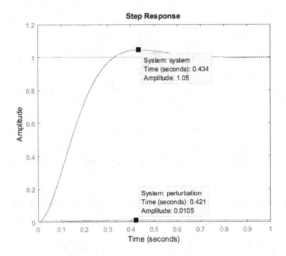

FIGURE 10.11: Response for the Hubble telescope.

If $K = 36$, $K_1 = 1.4 \times 6$ and $K/K_1 = 36/8.4$, and if $K = 100$, $K_1 = 1.4$, the step, the ramp, and the perturbation responses are shown in Figure 10.11 and can be obtained with the following script

```
% This is file Example10_9.m
% We analyze the behavior of the Hubble telescope
clear, clc, close all
% We first evaluate the plant
K = 100;
K1 = 14;
nump = [ 1 ];
denp = [ 1 K1 0 ];
% Transfer function for the plant
plant = tf( nump, denp )
% Transfer function from r(t)
system = feedback ( K*plant, [ 1 ] )
% Perturbation
% Transfer function from the perturbation
perturbation = feedback ( plant, [ K ] )
step(system)
hold on
step (perturbation)
hold off
```

We can see from Figure 10.11 that the specifications are satisfied for the design.

10.5 Applications to Chemical Engineering

Example 10.10 Mass Transfer

We wish to find the water vapor flux that evaporates from a vessel with water at 25° at a pressure of 1 atm. The distance from the water surface to the top of the vessel is 0.4 m. The water vapor flux is given by Bird's equation [1, 2]

$$N_z = -cD\frac{\partial x}{\partial z} - x\left(\frac{cD}{1-x_0}\right)\frac{\partial x}{\partial z}\bigg|_{z=0} \tag{10.15}$$

where N_z is the water vapor flux at a point z (the z axis is the vertical one), C is the total concentration in the vapor phase, D is the water diffusion coefficient in the air, x is the molar fraction of the water vapor, and x_0 is the value of x at $z = 0$. The non-steady concentration, assuming that there is no air flow in the vessel opening, can be obtained with

$$\frac{x}{x_0} = C = \frac{1 - \mathrm{erf}(Z - \varphi)}{1 + \mathrm{erf}\varphi} \tag{10.16}$$

where

$$Z = \frac{z}{\sqrt{4Dt}} \tag{10.17}$$

Here, t is time and φ can be found from the solution of the non linear equation: for N_z,

$$x_0 = \frac{1}{1 + [\sqrt{\pi}(1 + \mathrm{erf}\varphi)\varphi e^{\varphi^2}]^{-1}} \tag{10.18}$$

For the system we have the following constants:

$$D = 2.2 \times 10^{-5} \ \mathrm{m^2/s}$$

$$X_0 = \frac{P^{sat}(@25°C)}{Pi} = 0.0312$$

$$c = \frac{P_t}{RT} = 0.049 \ \mathrm{Kmol/m^3}$$

To solve this problem, first we find φ from equation (10.18). We then rewrite that equation as

$$\frac{1}{1 + [\sqrt{\pi}(1 + \mathrm{erf}\ \varphi)\varphi\ e^{\varphi^2}]^{-1}} - x_0 = 0$$

We can now create an m-function for this function as

```
% Non linear equation to solve for phi.
function phi_a = phi_b( x , x0)
global x0
phi_a = 1/(1 + (sqrt(pi)*( 1 + erf(x))*x*exp(x∧2))∧(-1))-x0;
```

The script to calculate the water vapor flux is:

```
% File Example10_10.m
% This file calculates the water vapor flux
% through the reservoir top.
close all, clc, clear
global x0
syms z
x0 = 0.0312;
z0 = 0;
c = 0.0409;
t = eps: 20 :3600;
z0 = 0.1; % Distance from surface of water to top of vessel.
D = 2.2e-5; % Water concentration.
x0 = 0.0312; % Molar fraction.
% Solution of the equation
phi = fzero('phi_b', 1);
x = x0*(1-erf(z./sqrt(4*D*t)))/(1 + erf(phi));
a = diff(x, z);
zi = [0.1 0.2 0.3 ];
cc = subs(a, z, 0);
for i = 1:3
    b = subs(a, z, zi(i));
    xx = subs (x, z, zi(i));
    N(i, :) = -c*D*b-c*D*xx.*cc/(1-x0);
end
plot( t/60, N*3600*18*1000 )
legend('z = 0.1', 'z = 0.2','z = 0.3')
xlabel('time(min)')
ylabel('N (gr/m2-hr)')
figure
for i = 1:3;
    dd(i, :) = subs(x, z, zi(i));
end
plot (t/60, dd(1,:), t/60, dd(2,:), '+', t/60, dd(3,:),'*')
legend('z = 0.1' , 'z = 0.3' ,'z = 0.3')
xlabel('time ( min)')
ylabel( 'Molar fraction for water' )
```

Figures 10.12 and 10.13 show the water vapor flux and the molar fraction.

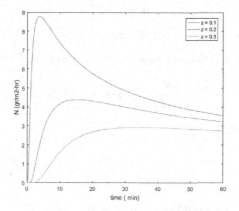

FIGURE 10.12: Flux versus time.

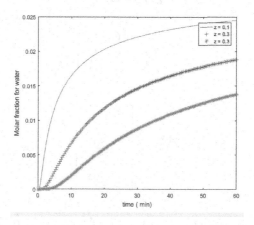

FIGURE 10.13: Molar fraction.

Example 10.11 Chemical reaction system
Let us consider the chemical reaction

$$A \xrightarrow{k_1,k_2} B \xrightarrow{k_3,k_4} C \qquad (10.19)$$

where the k_i are the kinetic rate constants. k_1 and k_3 are the kinetic rate constants to change $A \to B$ and $B \to C$ respectively, and k_2 and k_4 are the kinetic constants for $B \to A$ and $C \to B$, respectively. Let us assume the values for the kinetic constants are $k_1 = 1$ min^{-1}, $k_2 = 0$ min^{-1}, $k_3 = 2$ min^{-1}, $k_4 = 3$ min^{-1}. The initial concentrations at time $t = 0$ are:

$$C_A(0) = 1$$

$$C_B(0) = 1$$

$$C_C(0) = 1$$

The rate for each reaction is given by the system of differential equations:

$$\frac{dC_A}{dt} = -k_1 C_A + k_2 C_B$$

$$\frac{dC_B}{dt} = k_1 C_A - k_2 C_B - k_3 C_B + k_4 C_C \qquad (10.20)$$

$$\frac{dC_C}{dt} = k_3 C_B - k_4 C_C$$

In matrix form, these equations can be written as

$$\dot{C}(t) = kC(t) \qquad (10.21)$$

$$C(t) = \begin{bmatrix} C_A(t) \\ C_B(t) \\ C_C(t) \end{bmatrix} \qquad (10.22)$$

$$\dot{C}(t) = \begin{bmatrix} \frac{dC_A(t)}{dt} \\ \frac{dC_B(t)}{dt} \\ \frac{dC_C(t)}{dt} \end{bmatrix} \qquad (10.23)$$

$$K = \begin{bmatrix} -k_1 & k_2 & 0 \\ k_1 & -k_2 - k_3 & k_4 \\ 0 & k_3 & -k_4 \end{bmatrix} = \begin{bmatrix} -1 & 0 & 0 \\ 1 & -2 & 3 \\ 0 & 2 & -3 \end{bmatrix} \qquad (10.24)$$

To solve this system of differential equations we define them in a function as

```
function Cdot = dc(t, C)
global k
Cdot(1) = -k(1)*C(1) + k(2)*C(2) ;
Cdot(2) = +k(1)*C(1)-k(2)*C(2) - k(3)*C(2) + k(4)*C(3);
Cdot(3) = k(3)*C(2) - k(4)*C(3);
Cdot = Cdot';
```

With the following script we solve the equations and plot the concentration profiles:

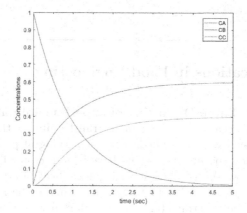

FIGURE 10.14: Concentration profiles for the components.

```
% This is file Example10_11.m
% It computes the component concentration profiles
% in a chemical reaction.
global k
% Input data:
fprintf( 'Enter the kinetic constants')
k(1) = input ('A->B , k1 = ');
k(2) = input ('B->A , k2 = ');
k(3) = input ('B->C , k3 = ');
k(4) = input ('C->B , k4 = ');
fprintf( 'Enter the initial concentrations' )
C0(1) = input('Initial concentration for A = ');
C0(2) = input('Initial concentration for B = ');
C0(3) = input('Initial concentration for C = ');
tmax = input( 'Enter the maximum time tmax = ');
tmax = 5;
k = [1, 0, 2, 3];
C0 = [1 0 0];
t = [ 0: 0.01: tmax ];
[t, C] = ode45('dc' , [0, tmax], C0);
plot(t, C)
xlabel(' tiempo (sec)')
ylabel('Concentrations')
legend('CA', 'CB', 'CC')
```

Figure 10.14 shows the component profiles. Note that at any time the principle of conservation of matter holds. That is,

$$C_A(0) + C_B(0) + C_C(0) = C_A(t) + C_B(t) + C_C(t) \qquad (10.25)$$

10.6   Applications in Food Engineering

Example 10.12 Three-D plotting of experimental data.
During the past decade, heat and mass transfer during the frying of meat products have been widely studied. The deep frying of pork meat has been studied and some experimental results have been presented[4]. In this paper, the authors have presented data for the specific heat in a tabular format. In this example we plot in a 3D plot the same data so the reader can have a better feeling of the experimental data. The data from the paper is tabulated and shown in Table 10.1. Here we see that there are four frying temperatures and four frying times. The script shown below realizes the 3D mesh plot. Figure 10.15 shows the 3D mesh plot and there we can see the behavior of the process.

TABLE 10.1:   Data for specific heat.

T(°C) (kJ/kg C)	t(min)	C_P
90	0	4.54
	30	3.56
	60	3.54
	90	3.36
	120	3.13
100	0	3.99
	30	3.78
	60	3.35
	90	3.31
	120	2.78
110	0	3.72
	30	3.58
	60	3.17
	90	2.85
	120	2.77
120	0	4.62
	30	4.43
	60	3.76
	90	3.46
	120	3.41

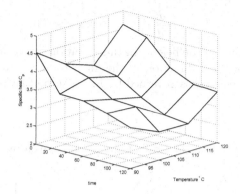

FIGURE 10.15: 3D plot for specific heat versus temperature and time.

```
% This is file Example10_12.m
% It plots the specific heat for pork meat
% in the frying process.
clear
close all
t = 0:30:120;
% Cp data at 90 degrees
Cp(1, :) = [4.54, 3.56, 3.54, 3.36, 3.13];
% Cp data at 100 degrees
Cp(2, :) = [3.99, 3.78, 3.35, 3.31, 2.78];
% Cp data at 110 degrees
Cp(3, :) = [3.72, 3.58, 3.17, 2.85, 2.77];
% Cp data at 120 degrees
Cp(4, :) = [4.62, 4.43, 3.76, 3.46, 3.41];
T = [90, 100, 110, 120]
[x, y] = meshgrid(t, T);
mesh(x, y, Cp)
axis([0 120 90 120 2 5])
xlabel('time')
ylabel('Temperature /\\circ C')
zlabel('Specific heat C_p')
```

Example 10.13 Fick's law.

Every diffusion process is governed by Fick's law. If the diffusion is in a steady-state, Fick's first law establishes that the concentration profile $c(x)$ and the diffusion flux J, which is the mass transported by unit area by unit time are related by

$$J = -D\frac{dC}{dx} \qquad \text{Fick's first law} \qquad (10.26)$$

where D is the diffusion coefficient. In this equation we assume that the diffusion takes place in the x direction. If the diffusion is not in the steady-state, then the diffusion process is governed by Fick's second law,

$$\frac{\partial C}{\partial t} = -D\frac{\partial^2 C}{\partial x^2} \qquad (10.27)$$

where $C = C(x,t)$. A solution for this equation is given by:

$$\frac{C_x - C_0}{C_s - C_0} = \frac{6}{\pi^2}\sum_{n=1}^{\infty}\frac{1}{n^2}exp\left(-n^2\frac{D\pi^2}{r^2}\right) \qquad (10.28)$$

where C_s is the surface concentration of the material diffusing, C_0 is the initial concentration, C_x is the concentration at a distance x from the surface, D is the diffusion coefficient, and r is the diffusion depth. In particular, we wish to know the humidity content of beet after being subjected to a drying process for a time t. The beet has been cut and shaped in spheres with a 2.5 cm diameter. The drying furnace has a temperature of 60°, the equilibrium humidity is 0.066 and the initial humidity is 0.25 kg_{water}/kg_{ss}. The beet diffusion coefficient is 7×10^{-10} m^2/s. The following m-file plots the humidity content during the first 60 seconds. In Figure 10.16 we plot the the solution in the series of Eq. 10.28, plotting from a single term in the series up to 13 terms. We see that this series converges very fast.

```
% This is file Example10_13.m
% Plots the drying process of beet
% according to Fick's law.
clear
clc
close all
y = zeros(11, 3601);
Xi = 0.25;
Xe = 0.066;
x1 = (6/pi^2);
t = [0:1:3600];
D = 7e-10;
r = 1.25e-2;
x = 0;
for n = 1: 13;
    x = x + x1*(1/n^2)* exp(-n^2*D*pi^2*t/r^2);
    X = x*(Xi-Xe) + Xe;
    y(n, :) = X;
end
```

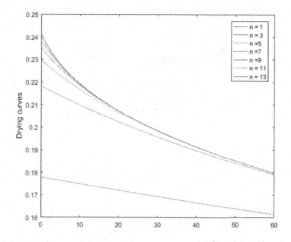

FIGURE 10.16: Drying curves for different number of terms in the series.

```
plot (t/60, y(1:2:13,:))
hold on
legend('n = 1','n = 3','n =5','n =7','n =9','n = 11','n = 13')
ylabel('Drying curves')
xlabel('time-minutes')
```

10.7 Applications in Civil Engineering

In this section we present two examples where MATLAB can be used. The first example is a beam deflection problem with a uniformly distributed load. The second problem is support reaction problem. Both examples are easily solved with MATLAB.

Example 10.14 Beam deflection

Let us consider a horizontal beam with length $L = 20$ m supported on both ends. It has a load uniformly distributed with weight $w = 100$ Kg/m. We wish to find the equation that describes the beam deflection.

Let us suppose that the left end is the origin of the coordinate system as shown in Figure 10.17. In the origin end, the beam has a forward upward vertical force given by $w \times L = 100 \times 20$ Kg. For any point P with coordinates (x, y) along the beam we have a load in the middle point of the segment OP given by wx. The differential equation is

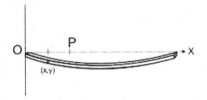

FIGURE 10.17: Deflected beam.

$$E \cdot I \cdot \frac{d^2y}{dx^2} = wLx - wx\left(\frac{1}{2}x\right) = wLx - \frac{1}{2}wx^2 \qquad (10.29)$$

where E is the elasticity modulus and I is the moment of inertia of a tranversal section. This differential equation can be solved in MATLAB by simply integrating twice with respect to x from $x = 0$ to $x = 20$. To integrate we can use the instruction **int** but we first rewrite the differential equation as

$$\frac{d^2y}{dx^2} = \frac{wLx}{EI} - \frac{wx}{EI}\left(\frac{1}{2}x\right) = w\left(Lx - \frac{x^2}{2}\right)\frac{1}{EI} \qquad (10.30)$$

In MATLAB we solve this as

```
d2y = w*(L*x - x^2/2)/(E*I)
```

The first integral is obtained with

```
dy = int(d2y)
```

and the second one with

```
dy = int(dy)
```

then we find the two constants of integration. If $E = 1000$, $I = 100$, $w = 200$ and $L = 10$. The complete script is

```
% This is file Example10_14.m
% This file integrates the differential equation
% for the moments.
% E is the elasticity model.
% I is the moment of inertia.
clc
clear
close all
syms x E I w L
% Differential equation.
```

```
d2y = w*(L*x - x^2/2)/(E*I);
% First integral,
dy = int(d2y);
y = int(dy);
% Evaluation of the constants of integration:
% y = 0 at x = 0; y = 0 at x = 20.
C2 = 0;
C1 = -w*L^3/(E*I)/3;
y = y + C1*x;
fprintf ('The solution is y = \n')
pretty(y)
x1 = [ 0:1:20 ];
% Substitution of normalized values for E and I.
% Substitution of w = 100, L = 10.
y1 = subs(y, [E, I, w, L], [1000, 100, 100, 10]);
y2 = subs(y1, 'x', x1); % Substitution of x by the vector x1.
% Maximum deflection at the center of the beam.
ymx = 5*w*L^4/( 24*E*I );
ymax = subs(ymx, [ E, I, w, L], [1000, 100, 100, 10 ])
plot ( x1, y2 )
```

After running the script we get

```
The solution is

y =


              3        4              3
    w (1/6 L x - 1/24 x ) - 1/3 w L x
    - - - - - - - - - - - - - - - - -   - - - - - - - -
              E I                            E I

ymax =
          2.0833
```

We also get the plot for the beam deflection shown in Figure 10.18. There we see the maximum beam deflection.

Example 10.15 Support reactions

Let us consider the support shown in Figure 10.18. To find out the reactions we use force equilibrium. The applied load is q_0. At point A, the only movement is a change in the angle θ and there are reactions at point A. At point C, there might be a movement that would change the coordinates (x, y) for point C, besides the reaction at point C as is shown in Figure 10.19. The forces and

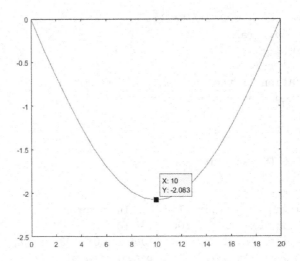

FIGURE 10.18: Beam deflection.

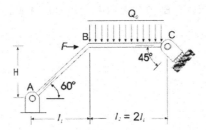

FIGURE 10.19: Support.

reactions present at the support are shown in Figure 10..20 and 10.21. The equilibrium equation at point A is

$$\sum M = 0 = -Fh - (2q_0L)(L_2) + R_e h \cos\theta_1 + R_{Ax} \qquad (10.31)$$

Furthermore, we have that $h = L_2 \tan\theta_2$. At point C, the components (x, y) for R_C are

$$\sum F_x = -F - R_C \cos\theta_1 + R_{Ax}$$
$$\sum F_y = 0 = R_{Ay} + R_C \sin\theta_1 - q_0 L_2 \qquad (10.32)$$

Using the instruction `solve` we can find the solutions. The following script produces the desired output:

```
% This is file Example10_15.m
% Evaluates the displacement of a structure.
clc, clear, close all
```

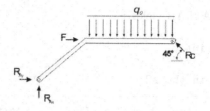

FIGURE 10.20: Reactions at the support.

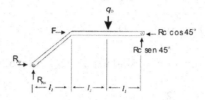

FIGURE 10.21: Components of the reactions.

```
syms L F q0 theta1 theta2 fi R rAx rax RAx RAy rAy L2
fprintf('Reaction at point C')
r = solve(-F*L*tan(theta2) - (2*q0*L*L2)...
    + R*cos(theta1)*L*tan(theta2) + R*sin(theta1)*3*L, R );
simplify(r);
expand(ans);
pretty(ans)
r1 = subs(r, theta2, pi/3);
r3 = subs(r1, theta1, pi/4);
r2 = subs(r3, L2, 2*L);
simplify(r2);
expand(ans);
fprintf('Reaction at C for theta1=60, theta2=45 and L2=2*L')
pretty(ans)
RAx = subs(rAx, theta1, pi/4);
expand(RAx);
simplify(ans);
fprintf('x reaction at A for theta1 = 60 and theta2 = 45')
pretty(ans)
ray = solve(rAy + r2*sin(theta1) - q0*L2, rAy);
ray1 = subs(ray, theta1, pi/4);
RAy = subs(ray1, L2, 2*L);
simplify(RAy);
expand(ans);
fprintf('y reaction at A for theta1 = 60 and theta2 = 45')
pretty(ans)
```

The output that MATLAB yields is

```
Reaction at point C

              F sin(theta2)
-----------------------------------------------------
cos(theta1) sin(theta2) + 3 sin(theta1) cos(theta2)

              q0 L2 cos(theta2)
+ 2 ----------------------------------------------------
     cos(theta1) sin(theta2) + 3 sin(theta1) cos(theta2)


Reaction at point C for theta1=60 and theta2=45 and L2=2*L

  1/2   1/2       1/2
 2    F3        2    q0 L
---------- + 4 -----------
   1/2              1/2
 3    + 3        3    + 3

x component at point A

   1/2    1/2   1/2              1/2
 -F3   - 3F  + 2   cos(theta1) F3   + 42 cos(theta1) q0 L
- - - - - - - - - - - - - - - - - - - - - - - - - - - - - - -
                   1/2
                 3    + 3

x component of reaction at point A for theta1=60 and theta2=45

  -3 F + 4 q0 L
  - - - - - - - -
    1/2
  3    + 3

y component for reaction at point A for theta1=60 and theta2=45

      1/2              1/2
    F3            q0 L            q0 L3
---------- + 2 --------- + 2 ---------
   1/2            1/2            1/2
 3    + 3       3    + 3       3    + 3
```

We can see that the output produced are expressions for the components for

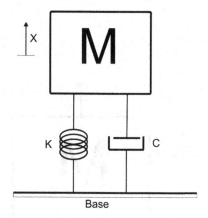

FIGURE 10.22: Mass-spring-damper system.

R_{Ax} and R_{Ay}. From here we can conclude that point C will move to the sides depending upon the magnitude of force F.

10.8 Applications in Mechanical Engineering

Example 10.16 Damped system under a harmonic movement at the base.

Sometimes, the base of a mass-spring-damper system is under a harmonic movement. Figure 10.22 shows the system and Figure 10.23 shows the effect of the harmonic movement at the base. The equation governing this system is

$$m\ddot{x} + c[\dot{x} - \dot{f}(t)] + k[x - f(t)] = 0 \qquad (10.33)$$

Since the force is given by $f(t) = F\sin(\omega t)$, the system equation becomes

$$m\ddot{x} + c\dot{x} + kx = kf(t) + c\dot{f}(t) \qquad (10.34)$$

To solve this equation, we rewrite the differential equation as a system of first order linear differential equations. Since

$$f(t) = F\sin(\omega t)$$
$$\dot{f}(t) = F\omega\cos(\omega t)$$

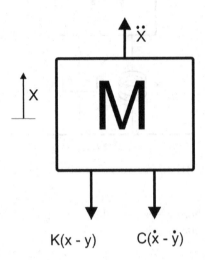

FIGURE 10.23: Effect of harmonic movement at the base.

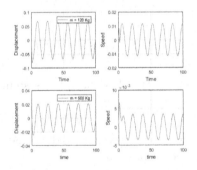

FIGURE 10.24: Displacement and speed plots.

$$\begin{cases} x_1 = x \\ x_2 = \dot{x} = \dot{x}_1 \\ \dot{x}_2 = \ddot{x} = -\frac{k}{m}x - \frac{c}{m}\dot{x} + \frac{c}{m}\dot{f}(t) + \frac{k}{m}f(t) \\ = -\frac{k}{m}x_1 - \frac{c}{m}x_2 + \frac{cF\omega}{m}\cos(t) + \frac{kF}{m}\sin(t) \end{cases}$$

In terms of x_1, x_2, we have

$$\begin{cases} \dot{x}_1 = x_2 \\ \dot{x}_2 = -\frac{k}{m}x_1 - \frac{c}{m}x_2 + \frac{cF\omega}{m}\cos(t) + \frac{kF}{m}\sin(t) \end{cases}$$

This system of equations can be programmed in a MATLAB function as

```
function xdot = ec_dif(t, x)
global k w m wn F c
xdot(1) = x(2);
xdot(2) = -(c/m)*x(2)-(k/m)*x(1) + (c*F/m)*cos(w*t) + ...
        (F*k/m)*sin(w*t);
xdot = xdot';
```

The script to solve the system for $m = 120$ Kg and for $m = 500$ Kg is:

```
% This is file Example10_16.m
% Evaluates the path and the position coordinates.
% Time from 0 to 15 seconds.
close all, clear, clc
global k w m wn F c
time = linspace(0, 100, 100);
% Initial conditions
x0 = [ 0 0 ]';
w = 5.81778; m = 120;
k = 40e1; wn = sqrt(k/m); F = 0.1; c = wn*m;
% Calling the function to solve the
% system of differential equations.
[t , x] = ode23('ec_dif', time, x0);
% Plot of displacement versus time
subplot(2, 2, 1)
plot(t , x( : , 2))
grid on
xlabel('Time'), ylabel('Displacement')
legend('m = 120 Kg')
```

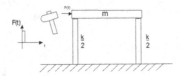

FIGURE 10.25: Structure excited by an impulse.

```
subplot(2, 2, 2)
plot(t , x ( : , 1) )
xlabel('Time'), ylabel('Speed')
grid on
m = 500; k = 40e1; wn = sqrt(k/m); c = wn*m;
[ t , x] = ode23('ec_dif', time , x0);
subplot(2,2,3)
plot( t, x( : , 2 ) )
legend('m = 500 Kg')
grid on
xlabel('time')
ylabel('Displacement')
subplot(2, 2, 4)
plot( t , x( : , 1) )
xlabel('time')
ylabel('Speed')
grid on
```

The plots produced are shown in Figure 10.24. The two plots at the left are the displacement and the ones to the right are the speed for the two different values of mass M. We readily see that when the mass is large, the displacement is small.

Example 10.17 Response of a structure to an impulse
To test the structure of Figure 10.25, we apply an impulse with a hammer. The magnitude of the impulse is F. If the system is underdamped the differential equation that controls its behavior is

$$m\ddot{x} + cx + k\dot{x} = 0 \qquad (10.35)$$

The initial conditions can be established by considering that for $t < 0$ we have that $x(0) = \dot{x}(0) = 0$. But at $t = 0^+$ we apply the impulse with a magnitude F. Since the momentum is conserved we then have that

$$\text{Impulse} * F = F\delta(t) = m\dot{x}(0^-) + m\dot{x}(0^+) = m\dot{x}(0^+) \qquad (10.36)$$

Thus,

$$\dot{x}(0^+) = \frac{F}{m} \qquad (10.37)$$

The displacement x at $t = 0^+$ has the same value it has at $x = 0^-$, that is,

$$x(0^-) = x(0^+) = 0$$

The differential equation can be rewritten as a set of linear first order differential equations as:

$$\begin{cases} \dot{x}_1 = x_2 \\ \dot{x}_2 = -\frac{k}{m}x_1 - \frac{c}{m}x_2 \end{cases}$$

This system of equations can be programmed in a MATLAB function as:

```
function xdot = ec_dif2( t, x)
global m k c f
xdot(1) = x(2);
xdot(2) = -(c/m)*x(2) - (k/m)*x(1);
xdot = xdot';
```

For values $m = 5$ Kg, $k = 2000$ N/m, $c = 10$ N s /m and $F = 10$ N, the following script plots the displacement, the speed, and a state plot of speed vs. displacement. The plots are shown in Figure 10.26.

```
% This is file Example10_17.m
% Evaluates the displacement and velocity of a structure
% under an impulse.
clear, clc, close all
global m k c f
% time from 0 to 7 seconds
time = linspace ( 0, 7, 1000);
% initial conditions x1 = 0, x2 = F/m .
x0 = [ 0 .2 ]';
% Data
m = 5; k = 2e3; c = 10;
% Calling the function to solve
% the system of differential equations
[t , x] = ode23('ec_dif2', time, x0);
% Plot of displacement
subplot(1, 3, 1)
plot( t, x( : , 1) )
grid on
xlabel('time')
ylabel('Displacement')
```

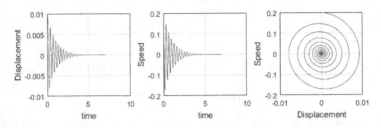

FIGURE 10.26: Plots of displacement and velocity vs. time and velocity vs. time.

```
% Plot of speed
subplot(1, 3, 2)
plot(t, x(:, 2))
grid on
xlabel('time')
ylabel('Speed')
% Plot of the path
subplot(1, 3, 3)
plot(x(:, 1), x(:, 2))
xlabel('Displacement')
ylabel('Speed')
grid on
```

10.9   Bibliography

1. J. Vélez-Ruiz y M.E Sosa Morales, Heat and mass transfer during the frying process of donuts, en Transport Phenomena in Food Processing, Eds. J. Welti-Chanes, J. Vélez-Ruiz y G. Barbosa, CRC Press Inc., Boca Ratón, Fl, 2003, pp. 429-444.

2. A. Soriano y J. Vélez, Evaluación de las propiedades físicas de los alimentos usando un programa de computadora, Información Tecnológica, Vol. 14, No. 4, pp. 23-28, La Serena, Chile, 2003.

3. R.B. Bird, et al. Transport Phenomena, J. Wiley and Sons, New York, 1960.

4. A. Constantinides and N. Mostoufi, Numerical Methods for Chemical Engineers with MATLAB Applications, Prentice-Hall PTR, Upper Saddle River, N.J., 1999.

5. M.E. Sosa-Morales, R. Orzuna-Espíritu, and J. Vélez-Ruiz, Mass, thermal, and quality aspects of deep-fat frying pork meat, Journal of Food Engineering,Vol. 77, pp. 731-738, 2006.

Chapter 11

MATLAB Applications to Physics

11.1   Introduction

The relationship between physics and mathematics is very strong. This importance has been recognized always and thus the Greeks tried to describe with mathematics every physical phenomenon they observed. Nowadays, every physics branch is described mathematically to be understood by peers and students. This chapter presents examples showing how MATLAB can be used to solve physics problems, taking advantage of the m-language and its graphics capabilities to display results. The examples presented in this chapter are on kinematics, dynamics, electromagetism, optics, astronomy, and modern physics.

11.2   Examples in Kinematics

The examples in this section cover topics related to a course in College Physics. The mathematics required here are as simple as solving an equation, differentiating, integrating, and solving differential equations. However, MATLAB provides very easy-to-use tools for displaying results in a numerical or graphical way. Our examples in this section include constant acceleration, free-fall, parabolic throw, and harmonic motion.

Example 11.1 Constant acceleration
The equations of motion for constant acceleration are given by the following equations [1]:

$$x = x_0 + v_0 t + \frac{1}{2}at^2 \tag{11.1}$$

$$v^2 = v_0^2 + 2a(x - x_0) \tag{11.2}$$

$$v = v_0 + at \tag{11.3}$$

$$x = x_0 + (v_0 + v)t \tag{11.4}$$

where x_0 and v_0 are the initial position and initial speed, respectively.

Now, a motorist is driving at a constant speed of 72 km/h on a 60 km/h road. A police officer who is standing next to the road spots the speeding car and at the moment the car passes beside it, he/she starts pursuing it with a constant acceleration of 4 m/s². We wish to know how long it will take the officer to catch up with the speeding car.

To solve this problem we have to find the time where both vehicles are together in the same position, that is, they have traveled the same distance. Since both vehicles have a constant acceleration, we can use the equations displayed above. We consider that at the time the speeding car passes besides the officer is position $x = 0$. If x_f and t_f are the position and time when the vehicles have traveled the same distance, then using Eq. 11.1 we have for the speeding car, which has an acceleration of 0 m/s²

$$x_f = 0 + 72t + \frac{1}{2}0t^2 \tag{11.5}$$

and for the police officer, which has no initial speed and a constant acceleration of 4 m/s²,

$$x_f = 0 + 0t + \frac{1}{2}4t^2 \tag{11.6}$$

These two simultaneous equations can be solved in MATLAB using the command `solve`, but first we have to convert the speeding car speed to m/s. Since 1 km = 1000 m and 1 hour = 3600 seconds the speed is then

$$v_{car} = \frac{72,000}{3600}\text{m/s} = 20\text{m/s} \tag{11.7}$$

The equations are then

$$x_f = 0 + 20t + \frac{1}{2}0t^2 \tag{11.8}$$

$$x_f = 0 + 0t + \frac{1}{2}4t^2$$

The instruction in MATLAB is then

```
[x, t] = solve('x = 20*t', 'x = 4*t^2/2')
```

which in MATLAB produces

```
[t, x] = solve('x = 20*t', 'x = 4*t^2/2')

    t =
      0
      10

    x =
      0
      200
```

We can see that there are two solutions and one of them is the trivial solution, that is, for t = 0 we get x = 0. The other solution gives t = 10 seconds and x = 200 meters. Then, it takes 200 m for the police officer to catch up to the speeding car. The time spent in the pursuit is 10 sec. We now can make a plot of the distance traveled by both vehicles. We can do this with

```
t = 0:0.01:12;
x_car = 20*t;
x_police = 4*t.^2/2;
plot( t, x_car, t, x_police)
xlabel('Time')
ylabel('Distance')
```

We show the plots in Figure 11.1. Now we plot the speed for both vehicles using Eq. 11.2. For the speeding car the speed is constant at 72 km/h. For the policeman we have the final speed is

$$v_{police} = \sqrt{v_0^2 + 2a(x - x_0)} = \sqrt{0 + 2*4*200} = 40 \text{ m/s}$$

A plot of speed for both vehicles can be plotted with

```
t = 0:0.01:12;
x_police = 4*t.^2/2;
v_police = sqrt(2*4*x_police);
v_car = 72;
plot(t, v_police, t, v_car)
xlabel('Time')
ylabel('Speed')
```

The resulting plot is shown in Figure 11.2.

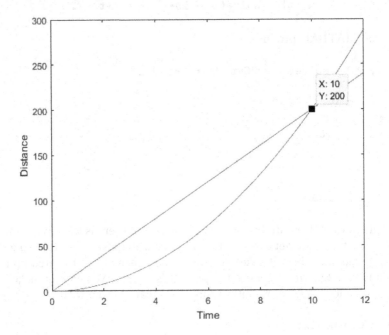

FIGURE 11.1: Distance traveled by both vehicles.

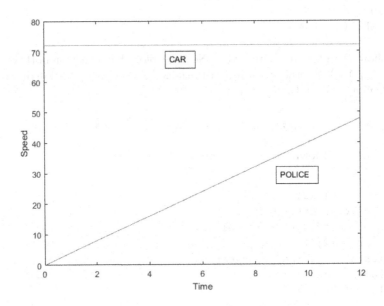

FIGURE 11.2: Plot of speed vs. time.

Example 11.2 Variable acceleration

Now suppose that the police car accelerates with an acceleration given by

$$a = t \text{ m/s}^2 \tag{11.9}$$

Since the acceleration is the derivative of the speed, then we have that

$$v_f = v_0 + \int a(t)dt = 0 + \int t\,dt = \frac{t^2}{2} \tag{11.10}$$

Also, the distance traveled by the police car is the integral of the speed. Then,

$$x = x + \int v(t)dt = 0 + \int \frac{t^2}{2}dt = \frac{t^3}{6} \tag{11.11}$$

For the speeding car the position is given as before by

$$x = 20\,t \tag{11.12}$$

We can solve again for position and time with MATLAB with

```
[t, x] = solve('x = 20*t', 'x = 4*t∧3/6')
    t =
        0
        30∧(1/2)
        -30∧(1/2)

    x =
        0
        20*30∧(1/2)
        -20*30∧(1/2)
```

MATLAB gives a symbolic solution that we can convert to a real number with

```
double([t, x])

    ans =
        0              0
        5.4772         109.5445
        -5.4772        -109.5445
```

We see that there are three solutions. This is because we have a cubic equation. Of the three solutions only the second one is a valid one. (The first solution is the trivial solution and the last one has negative time.) Thus, the time it takes the police car to reach the speeding car is $t = 5.4772$ seconds. Because the acceleration increases with time, the police car pursues the speeder for a shorter period of time.

Example 11.3 Free-falling ball

A baseball drops from a building roof 40 m height. We wish to know how long it will take to hit the sidewalk and what is the speed at the time of the impact. This is a constant acceleration movement and the value of acceleration is that due to gravity which has the value $g = 9.8$ m/s^2. The initial speed is $v_0 = 0$. Using Eqs. 11.3 and 11.4 we get the set of simultaneous equations

$$v = 0 - gt \tag{11.13a}$$

$$0 = 40 + vt \tag{11.13b}$$

We can solve this set of equations with

```
[t, v] = solve('v = -9.8*t', 'v*t + 40 = 0')
    t =
        -2.0203050891044214982881267488710
        2.0203050891044214982881267488710
    v =
        19.798989873223330683223642138936
        -19.798989873223330683223642138936
```

Again, this is a symbolic result which can be converted to a real one with:

```
double([t, v])
    ans =
        -2.0203      19.7990
        2.0203      -19.7990
```

The result with $t < 0$ is not a valid solution and, thus, the solution is

$$t = 2.0203 \text{ sec} \qquad v = \text{-19.799 m/s}$$

that is, the time it takes the ball to hit the floor is 2.0203 s and the speed when it hits the floor is 19.799 m/s.

Example 11.4 Parabolic throw

A projectile is thrown with an initial speed of $v_0 = 27$ m/s and with an angle $\theta = 57°$ with respect to the horizontal. Find the maximum distance that the projectile will travel before hitting the ground.

The speed components are

$$v_{ox} = 27 \cos \theta = 14.7053 \text{ and } v_{oy} = 27 \sin \theta = 22.6441$$

For the path coordinates (x, y), we have that the y component is affected by the gravitational acceleration g. The x component, ignoring air resistance, is not affected by gravity. Thus, the problem solution is found if we find the position of the ball when $y = 0$. To solve this problem we define first the acceleration components for the projectile. They are:

$$\ddot{x} = 0$$
$$\ddot{y} = -g$$

They can be written as a vector of first order differential as

$$\dot{x}_1(t) = x(t) \qquad \text{x-component of the ball position}$$
$$\dot{x}_2(t) = y(t) \qquad \text{y-component of the ball position}$$
$$\dot{x}_1(t) = \dot{x}(t) = x_3(t) \quad \text{x-component of the ball speed}$$
$$\dot{x}_2(t) = \dot{y}(t) = x_4(t) \quad \text{y-component of the ball speed}$$

We can rewrite these equations as

$$\dot{x}_1(t) = x_3(t)$$
$$\dot{x}_2(t) = x_4(t)$$
$$\dot{x}_3(t) = \ddot{x}(t) = 0$$
$$\dot{x}_4(t) = \ddot{y}(t) = -g = -9.8$$

The initial conditions for the variables are

$$x_1(t) = 0$$
$$x_2(t) = 0$$
$$x_3(t) = v_{x0} = 14.7053$$
$$x_4(t) = v_{y0} = 22.6441$$

This set of first order simultaneous differential equations can be solved if we first write them in an m-function as

```
function x_dot = projectile(t, x)
g = 9.8; % gravitational acceleration.
% Initialization of the vector of derivatives.
x_dot(1) = x(3);
x_dot(2) = x(4);
x_dot(3) = 0;
x_dot(4) = -9.8;
x_dot = x_dot';
```

This set of differential equations can be solved with the following script:

```
% This is file Example_11_4.m
% This script solves the equations of motion
% for the projectile throw and plots the path.
dt = 0: 0.001: 5;
```

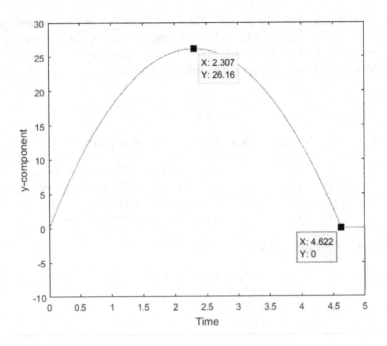

FIGURE 11.3: Plot of the y-coordinate for the position vs. time.

```
[t, x] = ode45('projectile', dt, [0, 0, 14.7053, 22.6441])
for n = 1: 5000;
   if x(n, 2)<= 0
      x(n, 2) = 0;
   end
end
plot(t, x(:, 2))
xlabel('Time')
ylabel('y-component')
```

The position is plotted in Figure 11.3 and there we see the maximum height
and the time when the ball falls to the ground.

11.3   Examples in Dynamics

In dynamics, the forces acting on the objects have to be taken into consideration. Our first example is the parabolic throw but this time considering the

effect of air resistance. Our second example is the movement of the pendulum, and our last example is a system composed of a mass, a spring, and a damper.

Example 11.5 Parabolic throw with air resistance

We here analyze the parabolic throw but this time we take into consideration air resistance. The air opposes to the movement of the ball with a force that is proportional to the speed squared as

$$\overrightarrow{F} = -cv^2 \frac{\overrightarrow{v}}{|\overrightarrow{v}|} \tag{11.14}$$

where c is a known constant. The equations of motion are now

$$m\ddot{x} = F_x = -c\dot{x}\sqrt{\dot{x}^2 + \dot{y}^2} \tag{11.15a}$$

$$m\ddot{y} = -mg + F_y = -mg - c\dot{y}\sqrt{\dot{x}^2 + \dot{y}^2} \tag{11.15b}$$

If we write these equations as a system of first order ordinary differential equations we have

$$\dot{x}_1(t) = x_3(t)$$
$$\dot{x}_2(t) = x_4(t)$$
$$\dot{x}_3(t) = \ddot{x}(t) = -\frac{c}{m}\dot{x}\sqrt{\dot{x}^2 + \dot{y}^2} = -\frac{c}{m}\dot{x}_3\sqrt{x_3^2 + x_4^2}$$
$$\dot{x}_4(t) = \ddot{y}(t) = -\frac{g}{m} - \frac{c}{m}\dot{x}_4\sqrt{x_3^2 + x_4^2}$$

This system can be programmed in a function as

```
function x_dot = projectile_air(t, x)
g = 9.8; % gravitational acceleration.
g = 9.8; % gravitation acceleration.
c = 0.001; % Constant in the air force equation.
m = 1; % mass of the ball.
% Initialization of the vector of derivatives.
x_dot(1) = x(3);
x_dot(2) = x(4);
x_dot(3) = -(c/m)*x(3)*sqrt(x(3)^2 + x(4)^2);
x_dot(4) = -g/m-(c/m)*x(4)*sqrt(x(3)^2 + x(4)^2);
x_dot = x_dot';
```

The following file solves the system of differential equations:

```
% This is file Example_11_5.m
% This script solves the motion equations for
% the projectile throw and plots the path when
% we consider the effect of air resistance.
dt = 0: 0.001: 5;
[t, x] = ode45('projectile_air', dt, [0, 0, 14.7053, 22.6441])
```

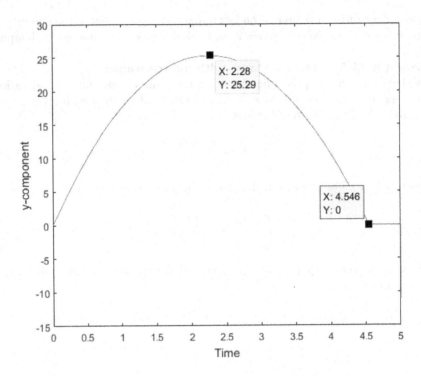

FIGURE 11.4: Plot of the y-coordinate for the position vs. time when air resistance is taken into account.

```
for n = 1: 5000;
    if x(n, 2)<= 0
        x(n, 2) = 0;
    end
end
plot(t, x(:, 2))
xlabel('Time')
ylabel('y-component')
```

The results of this script are shown in Figure 11.4. We readily see that the distance traveled by the ball has been reduced. The height is also less than the height when there is no air resistance.

Example 11.6 Simple pendulum

The pendulum is one of the most common examples in physics showing a harmonic movement. The pendulum is a mass m attached to a rod of length L of negligible mass and with a damping coefficient B. The angle θ is the angle between the vertical line and the rod. The equation of motion is

$$F_T = -w\sin\theta - BL\dot{\theta} \tag{11.16}$$

where F_T is the tangential force acting on the mass m, w is the weight given by $w = mg$, g is the gravitational acceleration. From Newton's second law we have

$$F_T = m\frac{dv}{dt} = mL\ddot{\theta} \tag{11.17}$$

Then, the equation of motion for the pendulum is

$$mL\ddot{\theta} + BL\dot{\theta} + w\sin\theta = 0 \tag{11.18}$$

This differential equation, for small values of θ, can be written as

$$mL\ddot{\theta} + BL\dot{\theta} + w\theta = 0 \tag{11.19}$$

because for small θ we have $\sin\theta \approx \theta$. This differential equation can be solved with dsolve. Assuming that $m = 2$ Kg, $L = 0.6$ m, and $B = 0.08$ Kg/m/s, we have $mL = 1.2$, $BL = 0.048$, and $w = 2(9.8) = 19.6$, and the differential equation is

$$1.2\theta + 0.048\dot{\theta} + 19.6\theta = 0$$

Now, if the initial position of the pendulum is

$$\theta = \pi/2 \qquad \dot{\theta} = 0$$

Using x for θ, the instruction dsolve is then

```
x=dsolve('1.2*D2x + 0.048*Dx + 19.6*x=0','x(0)=pi/2','Dx(0)=0')
```

which gives the solution

```
x = (pi*exp(-t/50)*cos((367491∧(1/2)*t)/150))/2 + ...
    (367491∧(1/2)*pi*exp(-t/50)*sin((367491∧(1/2)*t)/150))/244994
```

We now plot this solution with the instruction ezplot as

```
ezplot(x,[0, 10])
xlabel('Time')
ylabel('Angular position \theta')
```

The resulting plot is shown in Figure 11.5. We see how the damping factor makes the oscillation amplitude to decrease.

Example 11.7 Simple pendulum for large θ

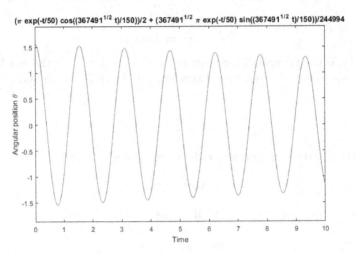

FIGURE 11.5: Plot of angular position vs. time.

Now we analyze the behavior of the pendulum when the oscillation is large, that is, the angle θ is large so the approximation $\sin\theta \ncong \theta$ is not valid. Thus, we use Eq. 11.14 repeated here for convenience:

$$mL\ddot{\theta} + BL\dot{\theta} + w\sin\theta = 0 \qquad (11.20)$$

This equation cannot be solved with the instruction `dsolve` because it cannot find a closed-form solution. We have to use numerical methods to find a solution. We first have to obtain a set of linear differential equations. We define $x_1(t) = \theta$ and $x_2(t) = \dot{\theta}$. Now, Eq. 11.16 can be written as

$$\dot{x}_1(t) = x(2) \qquad (11.21a)$$

$$\dot{x}_2(t) = -\frac{B}{m}x_2(t) - \frac{w}{mL}\sin[x_1(t)] \qquad (11.21b)$$

These equations are described in the file `pendulum.m` as:

```
function xdot = pendulum(t, x)
% It defines the equation for the pendulum movement.
g = 9.8; % gravitational acceleration
w = 2; % pendulum weight
L = 0.6; % Rod length
B = 0.08; %Damping factor
m = w/g; % Mass of the pendulum
% Initialization of the vector of derivatives.
xdot(1) = x(2);
xdot(2) = -B/m*x(2) - w/(m*L)*sin(x(1));
```

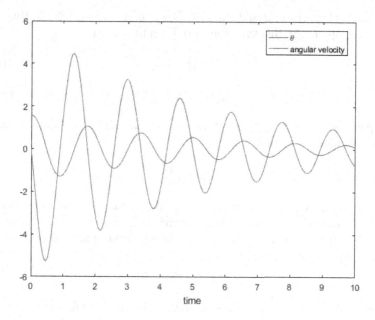

FIGURE 11.6: Plot of angular position and speed vs. time for the exact equations.

```
xdot = xdot';
```

Using the following file `Example_11_7.m` we obtain the plots for the solutions shown in Figure 11.6. This is a better approximation for the pendulum oscillations because we used the exact equation of motion.

```
% This is file Example_11_7.m
% It solves the equation for the pendulum problem.
t0 = 0; % Initial time.
tf = 10;
x0 = [pi/2 0]; % Initial conditions.
[t, x] = ode45('pendulum', [t0, tf], x0);
plot(t, x)
legend('\theta', 'angular speed')
xlabel('time')
```

Example 11.8 Parachute diving

Let us suppose a parachuter jumps in free fall from an airplane at $t = 0$ s. from a height h and opens the parachute at time t_0. From the initial jump up to time t_0 he/she goes in free fall with acceleration g. The speed in this time interval is $v(t) = gt$ and the elevation is $y(t) = h - gt^2/2$. At time t_0 the parachute opens. At this time the initial conditions are

$$v(t_0) = gt_0 \tag{11.22a}$$

$$y(t_0) = h - g\frac{t_0^2}{2} \tag{11.22b}$$

The parachuter must open in a time t such that $y(t) > 0$. That is, from Eq. 11.22b we must have that

$$h > g\frac{t_0^2}{2} \tag{11.23}$$

In other words, we must have $t_0 < 2h/g$. When the parachute is opened, the speed decrease due to air resistance which becomes evident as a drag force proportional to the speed squared and the equation of motion becomes

$$m\frac{dv}{dt} = -mg + kv^2 \tag{11.24}$$

where m is the parachuter mass and k is the drag coefficient (at sea level has the value 1.29 kg/m³) which is a function of the air density ρ and the parachute cross-sectional area A. It can be approximated as

$$k = 0.04\rho A \tag{11.25}$$

When the drag force equates the gravity, the parachuter reaches the maximum speed because then we have a vanishing acceleration. The equation of motion becomes

$$0 = -mg + kv^2$$

and the final speed is

$$v_f = \sqrt{\frac{mg}{k}}$$

To solve the equation of motion we write it as

$$\frac{dv}{dt} = -g + \frac{k}{m}v^2$$

and we describe it in the function:

```
function equation = parachuter(t, v)
rho = 1.29; % air density.
h = 2000; % height.
A = 10; % parachute cross-sectional area.
g = 9.8; % gravitational acceleration.
m = 80; % parachuter mass.
t0 = 15; % time when the parachute is opened.
```

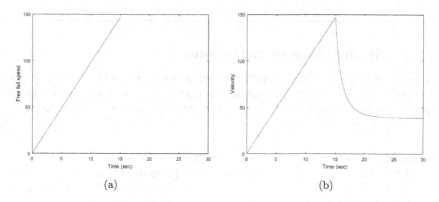

FIGURE 11.7: Parachute diving. a) Free fall, b) Parachute open.

```
ti = 0; % initial time.
k = 0.04*rho*A; % Drag coefficient.
equation = - (-g + (k/m)*v.*v);
```

We run this function with the file:

```
% This is file Example_11_8.m
t0 = 15;
t1 = linspace( 0, t0, 41);
tf = 30;
dt2 = linspace(t0, tf);
g = 9.8;
v1 = g*t1;
plot(t1, v1)
xlabel('Time (sec)')
ylabel('Free fall speed')
axis([0 30 0 150])
fprintf('Press enter to continue')
pause
hold on
v0 = v1(41);
[t, v] = ode45('parachuter', [t0, tf], v0);
plot(t, v)
ylabel('Velocity')
```

The results are shown in Figure 11.7 where we see the free fall speed from the beginning of the jump to the time when the parachute is opened. Then we see how the speed begins to decrease until it reaches a constant value.

11.4   Applications in Astronomy

Astronomy is an area where mathematics plays an important role and thus, MATLAB can find many applications. In this section we show how to plot the orbit of a planet around the Sun and how Mercury moves around the Sun as seen from Earth.

Example 11.9 Orbit of a planet around the Sun
To plot a planet's orbit around the Sun, we suppose that it has a mass m such that if M is the Sun's mass we have that $M >> m$. The force between the Sun and the planet is given by Newton's law of universal gravitation:

$$F = G\frac{Mm}{r^2} \tag{11.26}$$

The components of this force are

$$F_x = F\frac{x}{r} = G\frac{Mmx}{r^3} \tag{11.27a}$$

$$F_y = F\frac{y}{r} = G\frac{Mmy}{r^3} \tag{11.27b}$$

From Newton's second law

$$m\frac{dv_x}{dt} = -G\frac{Mmx}{r^3} \tag{11.28a}$$

$$m\frac{dv_y}{dt} = -G\frac{Mmy}{r^3} \tag{11.28b}$$

which can be rewritten as

$$\frac{dv_x}{dt} = -G\frac{Mx}{r^3} \tag{11.29a}$$

$$\frac{dv_y}{dt} = -G\frac{My}{r^3} \tag{11.29b}$$

Since we are only interested in finding the planet's orbit, we set $GM = 1$, we assume that the initial position is $(x,y) = (0.5,0)$, and that the initial speed has the components
$$v_x(0) = 0 \qquad v_y(0) = 1.63$$

If we set $v_x = \dot{x}$ and $v_y = \dot{y}$, the equations of motion are now

$$\ddot{x} = -\frac{x}{(x^2+y^2)^{3/2}} \tag{11.30a}$$

$$\ddot{y} = -\frac{y}{(x^2+y^2)^{3/2}} \tag{11.30b}$$

These equations can be solved by MATLAB if we write them as a set of linear ordinary differential equations as

$$\begin{aligned}
x_1 &= x \\
x_2 &= \dot{x} = \dot{x}_1 = v_x \\
x_3 &= y \\
x_4 &= \dot{y} = \dot{x}_3 = v_y
\end{aligned} \tag{11.31}$$

and finally, the system is

$$\dot{x}_1 = x_2 \tag{11.32a}$$

$$\dot{x}_2 = -\frac{x_1}{\left(x_1^2 + x_3^2\right)^{3/2}} \tag{11.32b}$$

$$\dot{x}_3 = x_4 \tag{11.32c}$$

$$\dot{x}_4 = -\frac{x_3}{\left(x_1^2 + x_3^2\right)^{3/2}} \tag{11.32d}$$

This system can be solved in MATLAB using the function:

```
function xdot = planet(t, x)
   xdot(1) = x(2);
   xdot(2) = -x(1)/(x(1)^2 + x(3)^2)^(3/2);
   xdot(3) = x(4);
   xdot(4) = -x(3)/(x(1)^2 + x(3)^2)^(3/2);
   xdot = xdot';
```

We run this function with the m-file:

```
% This is file Example_11_9.m
%
% This file evaluates the trajectory of a planet
% around the Sun.
x0 = [0.5, 0, 0, 1.63];
x = [0 0 0 0];
[t, x] = ode23('planet', [0, 4], x0);
plot(x(:, 1), x(:, 3))
grid on
text( -0.01, 0, 'X Sun')
```

The resulting trajectory is shown in Figure 11.9. We readily see that the orbit is an elliptical one.

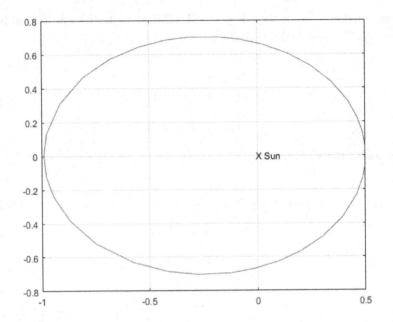

FIGURE 11.8: Orbit of a planet around the Sun.

Example 11.10 Mercury's orbit

Mercury has a very peculiar orbit around the Sun. It is described by the equations

$$x(t) = 93\cos t + 36\cos 4.15t \\ y(t) = 93\sin t + 36\sin 4.15t \tag{11.33}$$

These equations can be plotted with the following m-file:

```
% This is file Example_11_10.m
% This file plots Mercury's orbit around the Sun
% as seen from the Earth.
% Time:
t = [0: pi/360: 2*pi*22/3];
x = 93*cos(t) + 36*cos(4.15*t); % x-coordinate.
y = 93*sin(t) + 36*sin(4.15*t); % y-coordinate.
axis('square') % This makes a square plot.
plot(y, x)
axis('normal')
title('Mercury"s orbit seen from the Earth')
```

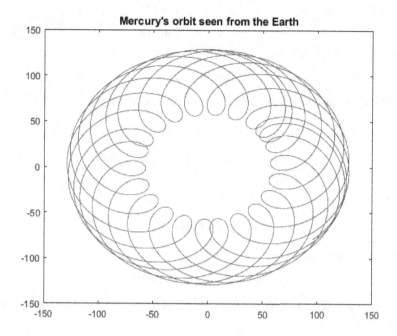

FIGURE 11.9: Mercury's orbit as seen from the Earth.

11.5 Applications in Electricity and Magnetism

The examples in this section plot the electrical field of a point charge and the magnetic field produced by a current flowing in a wire.

Example 11.11 Electric field produced by a charge point

The electric field due to a point charge is given by

$$\overrightarrow{\mathbf{E}}(r) = \frac{1}{4\pi\epsilon_0} \frac{Q}{r^2} \hat{\mathbf{r}} \tag{11.34}$$

where r is the distance from the point charge to the observation point. If we only wish to plot the field lines, we can normalize Eq. 11.30 with respect to $Q/4\pi\epsilon_0$. Then, Eq. 11.30 becomes

$$\overrightarrow{\mathbf{E}_n}(r) = \frac{\overrightarrow{\mathbf{E}}(r)}{\frac{Q}{4\pi\epsilon_0}} = \frac{1}{r^2}\hat{\mathbf{r}} = \frac{1}{\sqrt{x^2 + y^2}}\hat{\mathbf{r}} \tag{11.35}$$

where (x, y) are the coordinates of point r. The unit vector $\hat{\mathbf{r}}$ is given by

$$\hat{\mathbf{r}} = \hat{\mathbf{i}}\cos\theta + \hat{\mathbf{j}}\sin\theta \qquad (11.36)$$

where

$$\theta = \tan^{-1}\frac{y}{x}$$

and the normalized electric field components are given by

$$\vec{\mathbf{E}_n} = \left(\frac{1}{\sqrt{x^2+y^2}}\cos\theta, \frac{1}{\sqrt{x^2+y^2}}\sin\theta\right) \qquad (11.37)$$

To plot the field lines we first define a grid with x, y values:

```
[x, y] = meshgrid(-10: 2: 10);
```

Then, the instruction

```
quiver(x, y, Ex, Ey)
```

plots the field lines as arrows. Finally, we indicate the point charge position with

```
text(-0.01, 0, 'O')
```

The following m-file plots in Figure 11.10 the electric field lines:

```
% This is file Example_11_11.m
% This file plots the electric field lines
% of a point charge.
[x, y] = meshgrid(-5: 1.25: 5);
%
% Electric field equation
E = 1./(sqrt(x.∧2 + y.∧2));
% Unit vectors:
[unit_x] = cos(atan2(y, x));
[unit_y] = sin(atan2(y, x));
% Electric field components:
Ex = E.*unit_x;
Ey = E.*unit_y;
% plot
quiver(x, y, Ex, Ey)
text(-0.1, 0, 'O')
xlabel('x-axis')
ylabel('y-axis')
title('Electric field lines of a point charge')
```

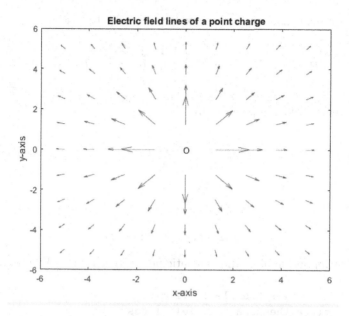

FIGURE 11.10: Electric field lines.

Example 11.12 Magnetic field lines

An electric current flowing in a wire generates a magnetic field. This discovery made by Oersted [2] led to a great deal of activity in electrodynamic research. Basically, if there is a current I flowing in a wire, a magnetic field is produced and it is given by

$$\overrightarrow{\mathbf{B}}(r) = \frac{\mu_0}{2\pi r^2} \overrightarrow{I} \times \overrightarrow{r} \qquad (11.38)$$

The magnetic field lines are concentric circles centered in the wire. In the case we have a wire coming out of the paper, we have magnetic field lines in planes parallel to the paper. If the x, y plane is on the paper and the z-axis is coming out of the paper, then the current is

$$\overrightarrow{I} = I\hat{\mathbf{k}} \qquad (11.39)$$

where $\hat{\mathbf{k}}$ is the z-axis unit vector. Since we are interested in the field lines we can normalize as we did in the previous example. Then

$$\overrightarrow{\mathbf{B}_n}(r) = \frac{1}{r^2} \overrightarrow{I} \times \overrightarrow{r} \qquad (11.40)$$

To plot the magnetic field lines we first need to make a grid with:

```
[x, y] = meshgrid(-20: 4: 20);
```

and then we calculate the vector product given by Eq. 11.36

$$\overrightarrow{B_n}(r) = \frac{1}{r^2}\overrightarrow{I} \times \overrightarrow{r} = \frac{1}{r^2}\begin{vmatrix} \hat{i} & \hat{j} & \hat{k} \\ 0 & 0 & I \\ x & y & 0 \end{vmatrix} = I(-y\hat{i} + x\hat{j})$$

The magnetic field components are

$$B_x = -\frac{y}{x^2 + y^2}$$

$$B_y = \frac{x}{x^2 + y^2}$$

use

$$\text{quiver(x, y, Bx, By)}$$

The following m-file plots the magnetic field lines shown in Figure 11.11:

```
% This is file Example_11_12.m
% It plots the magnetic field lines.
% Values for x,y.
[x, y] = meshgrid(-20: 4: 20);
Bx = -y./(x.^2 + y.^2);
By = x./(x.^2 + y.^2);
quiver(x, y, Bx, By)
title('Magnetic field lines')
```

11.6   Applications in Optics

Almost every phenomenon in optics is described by mathematical equations. In this chapter we plot the diffraction pattern of a two-slit screen. An interference pattern produced by two circular waves is plotted and there we can appreciate the maxima and minima in the interference pattern.

Example 11.13 Diffraction pattern of a two slit screen.

A front wave passes by a two slit screen and produces a diffraction pattern. If the slits have a width a, d is the distance between the slits, and λ is the light wavelength, the light intensity on a screen is given by

$$I(\theta) = I_m \cos^2 \beta \left(\frac{\sin \alpha}{\alpha}\right)^2 \tag{11.41}$$

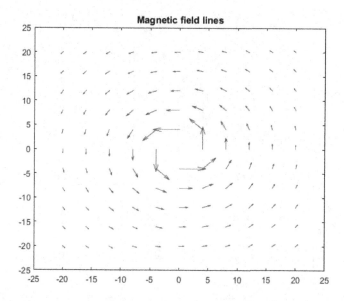

FIGURE 11.11: Magnetic field lines.

where

$$\beta = \frac{\pi d}{\lambda} \sin \theta \qquad \alpha = \frac{\pi a}{\lambda} \sin \theta$$

The following m-file plots the intensity on a screen when the slit separation is 50λ.

```
% This is file Example_11_13.m
% It plots the diffraction pattern for a two slit screen.
clear
close all
clc
Theta = [-pi/8:0.001: pi/8];
lambda = 1;
d = 50*lambda; % Slit separation.
a = [ lambda 5*lambda 7*lambda 10*lambda];
for k = 1 : 4;
    beta = pi*d*sin(Theta)/lambda;
    alpha = pi*a(k)*sin(Theta)/lambda;
    I = cos(beta).^2.*(sin(alpha)./alpha).^2;
    subplot(2, 2, k)
    plot(Theta, I)
    xlabel('Theta')
    ylabel('Intensity')
```

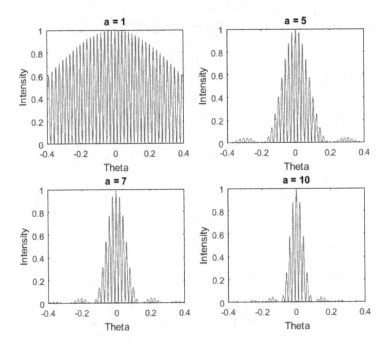

FIGURE 11.12: Diffraction patterns for a two slit screen when the slit distance is $d = 50\lambda$.

```
   if k == 1
      title('a = 1')
   elseif k == 2
      title('a = 5')
   elseif k == 3
      title('a = 7')
   else
      title('a = 10')
   end
end
```

The results are shown in Figure 11.12. Now, we change d to 6λ and the diffraction pattern changes as shown in Figure 11.13.

Example 11.14 Interference pattern
We can form an interference pattern by drawing two sets of circles. We can do that with the following m-file:

```
% This is File Example_11_14.m
% This file plots two families of circles to observe an
```

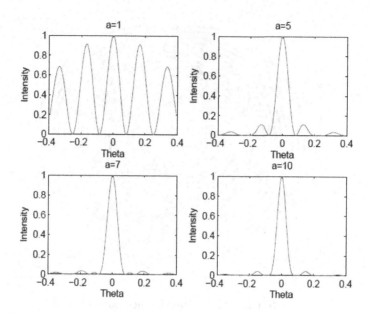

FIGURE 11.13: Diffraction patterns for a two slit screen when the slit distance is $d = 6\lambda$.

```
% interference pattern.
close all
N = 256;
t = (0:N)*2*pi/N;% Angle that goes from 0 to 2*pi radians.
hold on
r = 1:1:20 % Radii for the circles.
length(r)
for n = 1:length(r)
    % Plot of a family of circles.
    plot(r(n)*cos(t)-7, r(n)*sin(t), 'LineWidth', 3)
    % Plot of the second family of circles.
    plot(r(n)*cos(t), r(n)*sin(t), 'LineWidth', 3)
end
```

Figure 11.14 shows the plot and there we readily see an interference pattern.

11.7   Applications in Modern Physics

In this section we present two examples in Modern Physics. The first one uses equations from the Special Theory of Relativity and the second example com-

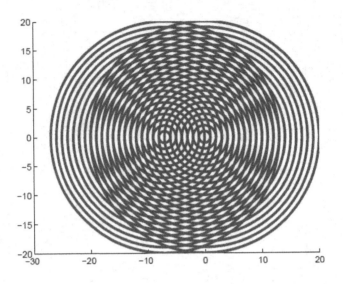

FIGURE 11.14: Interference pattern.

pares energy between Newtonian and modern physics.

Example 11.15 Time dilation
In phenomena taking place at speeds close to the speed of light c we have what is known as time dilation. In this case, if an event takes δt_M seconds in a moving frame, an observer in a reference frame has measured δt_R seconds. These time intervals are related by

$$\delta t_R = \frac{\delta t_m}{\sqrt{1 - v^2/c^2}} \tag{11.42}$$

This equation means that a clock at the reference stationary frame sees a longer time than a clock in the moving frame. Now, let us suppose that a twin brother travels at the speed $v = 0.9c$ and makes a round trip to a planet at a distance 2 light-years. How older is the twin brother when the traveler returns to Earth?

Because we have a round trip the total distance travelled is 4 light-years, where

$$d = 4 \text{ light-years} = 4 \times 3 \times 10^8 \text{m/s} \times \text{seconds in a year}$$

That is,

$$d = 4 \times 3 \times 10^8 \text{m/s} \times 3.1536 \times 10^6 \text{s} = 3.7843 \times 10^{16} \text{m}$$

For the traveller twin, the time elapsed is

$$\delta t_M = \frac{d}{0.9c} = \frac{3.7843 \times 10^{16}}{0.9 \times 3 \times 10^8} = 1.4016 \times 10^8 s = 4.4444 \text{ years}$$

On the other hand, for the twin brother on Earth, the time elapsed was

$$\delta t_R = \frac{\delta t_m}{\sqrt{1 - v^2/c^2}} = \frac{4.4444 \text{ years}}{\sqrt{1 - 0.9c/c}} = \frac{4.4444}{\sqrt{0.1}} = 14.0544 \text{years !!!}$$

Example 11.16 Classic and relativistic energies

The rest energy of an electron is defined by

$$E = mc^2 \tag{11.43}$$

which is the famous Einstein's equation for energy. According to Newtonian physics, when an electron is moving at speed v its kinetic energy is given by

$$K_N = \frac{1}{2}mv^2 \tag{11.44}$$

where m is the electron mass ($m = 9.109 \times 10^{-31}$ g). But according to modern physics, the kinetic energy is given by

$$K_M = \frac{mc^2}{\sqrt{1 - v^2/c^2}} - mc^2 \tag{11.45}$$

A plot of both energies is given in Figure 11.15. This plot is obtained with the following m-file:

```
% This is file Example_11_15.m
%It plots the kinetic energy of an electron moving at speed v.
m = 9.109e-31; % Electron mass.
c = 3e8; % Speed of light.
v1 = logspace(3, 8.95, 1000);
v = logspace(3, 8.47 ,1000);
Kn = m*v1.^2/2; % Newtonian energy.
Km = m*c^2./sqrt(1-v.^2/c^2)-m*c^2; % Relativistic energy.
line([c, c], [0, 4e-13])%Vertical line at the speed of light.
plot(v1,Kn,':',v,Km,'LineWidth',3)
title('Comparison between Newtonian and relativistic energies')
xlabel('Speed'), ylabel('Energy')
legend('Relativistic energy','Newtonian energy')
```

The results are shown in Figure 11.15. We readily see how both formulations give the same results for low speeds but as we approach the speed of light the Newtonian speed yields inaccurate results.

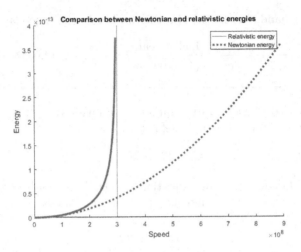

FIGURE 11.15: Comparison between Newtonian and relativistic energies.

11.8   Concluding Remarks

Physics makes extensive use of mathematics. Thus, many problems are amenable to be solved with a tool as efficient as MATLAB. We have presented problems in kinematics, dynamics, astronomy, optics, and modern physics. Of course, this set is by no means exhaustive but it gives a clear idea of what can be accomplished with MATLAB as a tool to aid in the solution of physics' problems.

11.9   Bibliography

1. H. D. Young and R. A. Friedman, University Physics, 11th Ed., Addison Wesley, San Francisco, 2004.

2. A. Beiser, Concepts of Modern Physics, 6th Ed., McGraw-Hill Book Co., New York, 2002.

3. R. A. Serway, C. J. Moses, and C. A. Moyer, Modern Physics, Brooks/Cole, Florence, KY, 2004.

4. F. Jenkins and H. White, Fundamentals of Optics, 4th Ed., McGraw-Hill Book Co., New York, 2001.

Chapter 12

Applications to Finance

12.1 Introduction

Finance, similar to engineering, makes extensive use of mathematics. In this chapter we present several financial problems showing how MATLAB can be successfully used to solve them. We start with examples that can be directly solved by MATLAB. Then, we show the use of some functions within the Financial Toolbox and the Financial Derivatives Toolbox. The examples included give the reader an idea of MATLAB's potential to solve problems in Finance. We start this chapter with a brief review of financial concepts and use MATLAB to solve some simple financial problems.

12.2 The Financial and Financial Derivatives Toolboxes

MATLAB has developed toolboxes to solve financial problems. These toolboxes are the Financial Toolbox and the Financial Derivatives Toolbox. Together, they group more than 200 functions that allow us to perform financial computations very easily. In this section, we will only describe a few of them by way of examples so the reader can see the potential of these two toolboxes.

Example 12.1 Bonds with fixed yield
Let us consider a zero coupon bond with a nominal value F, a one year maturity and a price P. The yield is R where

$$R = \frac{F}{P} \tag{12.1}$$

The yield rate is

$$r = R - 1 = \frac{F}{P} - 1 \tag{12.2}$$

which can be written as

$$P = \frac{F}{1+r} \tag{12.3}$$

If F and r are fixed, Eq. (12.3) can be seen as a discount formula. In the case of a cash flow we have a series of periodic payments P_t in discrete times $t = 0, 1, 2,..., n$. The present value of the cash flow is

$$VP = \sum_{t=0}^{n} \frac{C_t}{(1+r)^t} \tag{12.4}$$

If the frequency of the payments is larger, let us say there are m payments per year at regular intervals of time, the equation is changed to

$$VP = \sum_{k=0}^{n} \frac{C_t}{(1+r/m)^k} \tag{12.5}$$

where k represents the changes for each period and n is the number of years times the number of payments each year. Now, let us consider the case of a bond with a series of payments c_t $(t = 1, 2, ..., n)$. The return yield is defined as the value that makes the present value equal to zero, that is,

$$\sum_{t=0}^{n} \frac{C_t}{(1+\rho)^t} = 0 \tag{12.6}$$

With a change of variable $h = 1/(1 + r)$, the equation can be written as

$$\sum_{t=0}^{n} C_t h^t = 0 \tag{12.7}$$

To find the value of ρ we solve this equation in MATLAB with the function roots. If the cash flow is [-100 8 8 8 8 108], a complete procedure is

```
cf = [ -100 8 8 8 8 108 ]; % cash flow.
cf = fliplr(cf); % Flip data from left to right
h = roots(cf) % Root calculation.
rho = 1./h-1
```

and MATLAB produces

```
h =
-0.8090 + 0.5878i
-0.8090 - 0.5878i
0.3090 + 0.9511i
0.3090 - 0.9511i
0.9259

rho =
-1.8090 - 0.5878i
-1.8090 + 0.5878i
-0.6910 - 0.9511i
-0.6910 + 0.9511i
0.0800
```

Of the five roots, four of them are complex and one is real. The real root gives the rate of return as $\rho = 0.08$. All the above calculations made to find out the rate of return can be easily replaced with the instruction **irr** in the Financial toolbox as follows:

```
>> cf = [ -100 8 8 8 8 108 ]; % cash flow.
>> ro = irr( cf )

ro =
    0.0800
```

We readily see the power of the Financial toolbox functions to solve problems in finance.

Example 12.2 Present value
The instruction **pvvar** calculates the present value of a series with a discount rate. For example, we can calculate the value of a 5 year bond with a nominal value of 100 and a coupon rate of 8 % and 9 %. Starting with cf = [0 8 8 8 8 108] we have

```
>> cf = [ 0 8 8 8 8 108 ] ;
>> pvvar (cf, 0.08) % rate of 8 %.
```

To obtain:

```
ans =
    100.0000
```

If we now repeat with a rate of 9%,

>> pvvar (cf, 0.09) % rate of 9 %.

we obtain:

ans =
    96.1103

Note that the cash flow has a zero in the first position because the coupon is received at the end of the first year.

Example 12.3 Quality of the price change
Given a cash flow series happening at times $t_1, t_2, \ldots, t_n$, the duration of the series is defined as

$$VP = \frac{\displaystyle\sum_{k=1}^{n} \frac{k}{m} \frac{c_k}{(1+\lambda/m)}}{\displaystyle\sum_{k=1}^{n} \frac{c_k}{(1+\lambda/m)^k}} \qquad (12.8)$$

$$VP = \frac{\displaystyle\sum_{t=0}^{n} PV(t_n)}{PV} \qquad (12.9)$$

where PV is the present value of the series and $PV(t_i)$ is the present value of the cash flow c_i at time t_i, $i = 1, 2, \ldots, n$. For the case of a zero coupon bond which is a cash flow, the duration is simply the time to maturity. When we consider a generic bond we use the yield as the discount rate in the computation of the present value to obtain the Macauley duration, assuming m deposits per annum:

$$D = \frac{\displaystyle\sum_{k=1}^{n} \frac{k}{m} \frac{c_k}{(1+\lambda/m)}}{\displaystyle\sum_{k=1}^{n} \frac{c_k}{(1+\lambda/m)^k}} \qquad (12.10)$$

The derivative of the price with respect to the yield is given by

$$\frac{dP}{d\lambda} = \frac{d}{d\lambda}\left(\sum_{k=1}^{n}\frac{c_k}{(1+\lambda/m)^k}\right) = -\sum_{k=1}^{n}\frac{k}{m}\frac{c_k}{(1+\lambda/m)^{k+1}} \qquad (12.11)$$

We now define the modified duration as $D_M = D/(1+\lambda/m)$, so

$$\frac{dP}{d\lambda} = -D_M P \qquad (12.12)$$

We see that the modified duration is related with the slope of the curve price-yield. A better approximation is obtained with the convexity defined by

$$C = \frac{1}{P}\frac{d^2P}{d\lambda^2} \qquad (12.13)$$

For a bond with m coupons per year is

$$C = \frac{1}{P(1+\lambda/m)}\sum_{k=1}^{n}\frac{k(k+1)}{m^2}\frac{c_k}{(1+\lambda/m)^k} \qquad (12.14)$$

Using the modified duration and the convexity we can write an approximation to P as

$$\delta P \approx D_M P \delta\lambda + \frac{PC}{2}(\delta\lambda)^2 \qquad (12.15)$$

Now let us assume that we have a chain of four cash flows (10, 7, 9, 12) that happen at times $t = 1, 2, 3, 4$. We can calculate the present value of this series with different yields by using the MATLAB function pvvar:

```
>> cash_flow = [ 10 7 9 12 ]
>> % Calculate the present value with a rate of 5 %.
>> p1 = pvvar( [ 0, cash_flow ], 0.05)

p1 =
    33.5200

>> % Calculate the present value with a rate of 5.5%.
>> p2 = pvvar( [ 0, cash_flow ], 0.055)

p2 =
```

```
33.1190

>> p2 - p1

ans =
      -0.4010
```

In this example we have added a 0 in the first position of the cash flow series because the function **pvvar** assumes that the first cash flow happens in time $t = 0$. We see that an increase in the yield of 0.005 produces a drop in the price of 0.4010. Now we can calculate the **duration** and **convexity** with the MATLAB functions **cfdur** and **cfconv**:

```
>> cash_flow = [ 10 7 9 12 ];
>> [ d1 dm ] = cfdur (cash_flow , 0.05)

d1 =

    2.5369

dm =

      2.4161

>> convexity = cfconv (cash_flow, 0.05 )

convexity =
          9.4136

>> First_order = -dm*p1*0.005

First_order =
            -0.4049

>> Second_order = -dm*p1*0.005 + 0.5*convexity*p1*0.005^2

Second_order =
            -0.4010
```

which is the same result obtained above for **p2-p1**. We see that for small changes in the yield, the first order approximation is adequate but, the second order approximation is practically exact.

Example 12.4 Fixed income bonds

We recall from Ch. 3 the way date and time are formatted in MATLAB. For this we use `datestr(today)` and `datestr(now)` to obtain:

```
>> datestr(today)

ans =
    12-Feb-2008

>> datestr(now)

ans =
    12-Feb-2008 11:23:30
```

In yield calculations we need these functions. To calculate the `yield` of a bond we can use the MATLAB function `bndprice`. The format of this function is

```
[Price, Accrued_Interest] = bndprice(Yield, Coupon rate, ...
    Settlement date, Maturity)
```

`Price` is the clean price to which we must add the `Yield` to obtain the real price (dirty price). For example, for a `Yield` of 8% and a `Coupon rate` of 10%, a `Settlement date` 10-August-2008, and a `maturity` date 31-December-2020, we have:

```
>> [Price, Accrued_interest] = bndprice(0.08, 0.1, ...
        '10-aug-2008', '31-dec-2020')
```

To obtain:

```
Price =
     115.5228

Accrued_interest =
            1.1141

>> Price + Accrued_interest

ans =
     116.6370
```

Now, let us consider a vector of yields such as

```
yield = [0.01 : 0.001 : 0.2 ];
```

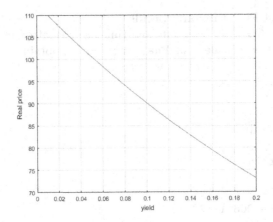

FIGURE 12.1: Plot of `Real price` vs. `yield`.

Repeating the calculation using `bndprice` with a coupon rate of 5% we obtain a vector `Price` and a vector `Accrued_interest`. With them we can plot the `real price` with the following script:

```
yield = [0.01 : 0.001 : 0.2 ];
[Price, Accrued_interest] = bndprice(yield, 0.05, ...
        '10-aug-2018', '31-dec-2020');
plot(yield, Price + Accrued_interest)
grid on
xlabel('yield')
ylabel('Real price')
```

to produce the plot of Figure 12.1. From this plot we can readily see that as the `yield` increases, the `Real price` decreases.

12.3 The Financial Derivatives Toolbox

In finance, the term derivative refers to a collection of assets falling into three broad categories, namely, options, future contracts, and swaps. A derivative then is a financial instrument whose value depends on the values of other financial instruments such as the price of a stock or of a traded asset [3]. MATLAB has developed the `Financial Derivatives Toolbox` which can be used to analyze individual derivative instruments and portfolios for several types of interest rate based instruments and equity based financial instruments. This toolbox contains a demo GUI that illustrates some of the functions that can be analyzed. This demo tool is called `derivtool` and is run by typing its name

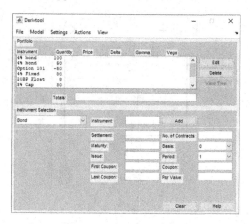

FIGURE 12.2: Window for **derivtool** with some preloaded data.

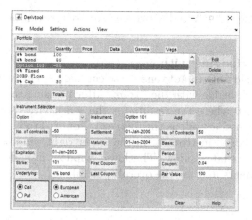

FIGURE 12.3: **Option 101** selected in **derivtool**.

in the MATLAB workspace. The GUI for **derivtool** is shown in Figure 12.2. As can be seen from the figure, **derivtool** comes with preloaded data. There is a **bond**, an **option**, a **fixed**, a **float**, a **cap**, a **floor**, and a **swap**. Figure 12.3 shows the tool with the **Option** selected. In the **Model** menu, the user can choose two models: **HJM** or **Zero Curve**. For the **Settings** we have **Initial Curve**, **Sensitivities**, **Volatility Model**, and **Tree Construction**. The actions that can be done are **Price** and **Hedge**. We can **View** the **Spot Rates** and **Unit Bond Prices**. For example, in the case of the **Option 101** we can calculate the **Sensitivities** by using **Settings** → **Sensitivities**. Doing this opens the window of Figure 12.4 and there we select the **Vega** sensitivity. After pressing **OK** we get the values of this sensitivity for all the instruments in the **Portfolio** window, as shown in Figure 12.5.

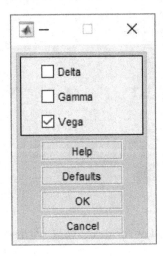

FIGURE 12.4: Window where sensitivities can be chosen.

Example 12.5 Hedging with `derivtool`
We can also do hedging. In `Actions` we select `Hedge`. This will do for the 20BP
`Float` and we will hedge with the `Floor`. After pressing the button `Hedge` we
obtain Figure 12.6 where the results for the sensitivities for the hedged and
hedging instruments, as well as the `Overall Portfolio` are shown. This window shows the hedged instrument in the upper right window and the hedging
instrument below that window. The lower window provides the sensitivities
for both instruments and for the overall portfolio.

12.4   The Black-Scholes Analysis

In 1973, Fischer Black and Myron Scholes published a paper where they derived a differential equation that applies to any derivative that is dependent
upon a non-dividend stock [1]. This differential equation can be solved to obtain values for European call and put options on the stock [3]. The so-called
Black-Scholes differential equation is

$$\frac{\partial f}{\partial t} + rS\frac{\partial f}{\partial S} + \frac{1}{2}\sigma^2 S^2 \frac{\partial^2 f}{\partial S^2} = rf \qquad (12.16)$$

Here, f is the function that satisfies the differential equation. In our case we
are looking for the Call option and the Put option. Additionally, S is the stock
price, T is the maturity period, r is the risk free interest rate, and σ is the
volatility. The solution depends upon the boundary conditions that are used.

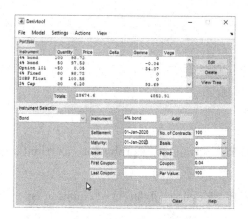

FIGURE 12.5: Values for **Vega** sensitivity are shown.

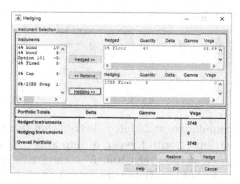

FIGURE 12.6: Hedging in `derivtool`.

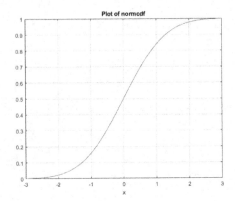

FIGURE 12.7: Cumulative probability distribution function.

A solution to the Black-Scholes differential equation is:

For the call option

$$c = S_0 N(d_1) K e^{-rT} N(d_2) \qquad (12.17)$$

and for the put option

$$p = K e^{-rT} N(-d_2) - S_0 N(-d_1) \qquad (12.18)$$

where

$$d_1 = \frac{ln(\frac{S_0}{K}) + (\frac{r+\sigma^2}{2})T}{\sigma\sqrt{T}} \qquad (12.19)$$

$$d_2 = \frac{ln(\frac{S_0}{K}) + (\frac{r-\sigma^2}{2})T}{\sigma\sqrt{T}} = d_1 - \sigma\sqrt{T} \qquad (12.20)$$

Here $N(x)$ is the cumulative probability distribution function for a variable that is normally distributed with a mean of zero and a standard deviation of 1, S_0 is the stock price at time zero, and K is the strike price. The function $N(x)$ is integrated into MATLAB as normcdf(x). A plot of it from x = -3 to x = +3 is shown in Figure 12.7.

The Financial Derivatives Toolbox has the function blsprice that computes the solution of the Black-Scholes equation. The format for this function is:

```
[Call, Put] = blsprice(Price, Strike,Rate,Time,Volatility,Yield)
```

An example shows the use of this instruction.

Example 12.6 European stock price
We wish to consider the situation in six months from the expiration of an option valued at \$42, the exercise price is \$40, the risk free interest rate is 10% per annum, and the volatility is 20% per annum[1]. The variables are then, $S = 42$, $X = 40$, $r = 0.1$, $\sigma = 0.2$, and $T = 0.5$. Running the MATLAB function `blsprice`

```
>> [Call, Put] = blsprice (42, 40, 0.1, 0.5, 0.2)
```

produces the `call` and `put` options as

```
Call =
     4.7594

Put =
     0.8086
```

This means that the stock price has to rise by \$2.76 for the purchaser of the call to break even. In a similar way, the stock price has to fall by \$2.81 for the purchaser of the put to break even.

We can plot the `Call` and `Put` options for several pair of variables. For example, if we are interested in observing the way the put option varies when the interest rate and time change, we can use a mesh plot. The following m-file does the required plot

```
r = 0.1: (.5-.1)/100: 0.5;
T = 0.1: (5-.1)/100: 5;
for i = 1:length(T)
    for j = 1:length(r)
        [c(i, j), p(i, j)] = blsprice(42, 40, r(j), T(i), 0.2);
    end
end
figure
surf(r,T,p)
xlabel('Interest rate')
ylabel('Time')
zlabel('Put option')
```

In the plot of Figure 12.8 we can observe that the put option begins to increase

---

[1]This example is taken from [2] Example 12.7, with permission of the author.

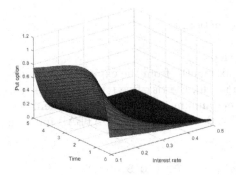

FIGURE 12.8: Variation of the put option with the interest rate and time.

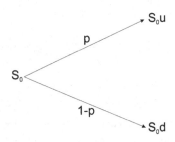

FIGURE 12.9: Binomial stock price movement.

with time but that at about $t = 1.5$ it begins to decrease.

12.4.1   American Options

An American option is an option that can be exercised at any time during the lifespan of the option. Unfortunately, there is no closed form solution to the problem as in the case of European options. Thus, the only way is to proceed with computer methods. The most popular numerical procedures are the use of trees, finite differences methods, and Monte Carlo simulations [3, 6]. The use of trees is a method devised by Cox, Ross, and Rubinstein [3] known as the CRR method.

Binomial trees are studied in many textbooks. The procedure starts by dividing the life of an option into a large number of small time intervals t. In each time interval the stock price moves from its initial value S_0 to one of two new values $S_0 u$ and $S_0 d$. In general, $u > 1$ and $d < 1$. The movement to $S_0 u$ is an up movement and the movement to $S_0 d$ is a down movement. The probability of an up movement is denoted by p and the probability of a down movement is then $1 - p$. The movement of the stock is shown in Figure 12.9.

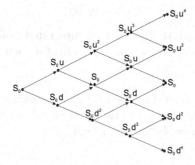

FIGURE 12.10: Complete binomial tree.

The parameters u, d, and p are given by

$$u = e^{\sigma\sqrt{\delta t}} \tag{12.21}$$

$$d = e^{-\sigma\sqrt{\delta t}} \tag{12.22}$$

$$p = \frac{a - d}{u - d} \tag{12.23}$$

where a is

$$a = e^{\sigma \delta t} \tag{12.24}$$

The complete tree of stock prices will look as shown in Figure 12.10.

Example 12.7 Binomial tree for an American option

Consider a five-month American put option on a non-dividend-paying stock which has a $50 price, the strike price is $50, the risk free interest rate is 10% per annum, and the volatility is 40% per annum[2]. That is, $S = 50$, $K = 50$, $r = 0.1$, $\sigma = 0.4$, and $T = 5/12 = 0.4167$. If the time interval is a month, then $t = 0.0833$. The remaining parameters are calculated from Eqs. (12.25) and (12.29) as

$$u = e^{\sigma\sqrt{\delta t}} = 1.1224 \tag{12.25}$$

$$d = e^{-\sigma\sqrt{\delta t}} = 0.8909 \tag{12.26}$$

$$a = e^{r\delta t} = 1.0084 \tag{12.27}$$

$$p = \frac{a - d}{u - d} = 0.5073 \tag{12.28}$$

---

[2]This example is taken from [2] Example 18.1, with permission of the author.

$$1 - p = 0.4927 \qquad (12.29)$$

The MATLAB Financial Toolbox has an integrated function that calculates the values of the nodes of a binomial tree. This function is binprice and has the general form:

```
[AssetPrice, OptionValue] = binprice(Stockprice, ...
        Strikeprice, Rate, Time, Increment, Volatility, ...
        Flag, DividendRate, Dividend, ExDiv)
```

In this function, flag = 1 for a call option and flag = 0 for a put option. DividendRate, Dividend, and ExDiv are optional parameters and are not used for a non-dividend-paying stock. For a tree with six steps, we have an increment of $(5/12)/6 = 0.0694$. Thus, for a put option

```
>> [Stockprice, Optionprice] = binprice(50, 50, 0.1, 5/12, ...
        0.0694, 0.4, 0, 0, 0, 0)
```

Stockprice =

50.0000	55.5583	61.7344	68.5971	76.2228	84.6961	94.1113
0	44.9978	50.0000	55.5583	61.7344	68.5971	76.2228
0	0	40.4961	44.9978	50.0000	55.5583	61.7344
0	0	0	36.4447	40.4961	44.9978	50.0000
0	0	0	0	32.7986	36.4447	40.4961
0	0	0	0	0	29.5173	32.7986
0	0	0	0	0	0	26.5643

Optionprice =

4.1745	2.0055	0.5882	0	0	0	0
0	6.4610	3.4894	1.2006	0	0	0
0	0	9.6043	5.8893	2.4507	0	0
0	0	0	13.5553	9.5039	5.0022	0
0	0	0	0	17.2014	13.5553	9.5039
0	0	0	0	0	20.4827	17.2014
0	0	0	0	0	0	23.4357

Since we set the flag = 0 the option price is for a put option. From the MATLAB output we see that the put option price is 4.1745.

For a call option the instruction is

```
>> [Stockprice, Optionprice] = binprice(50, 50, 0.1, 5/12, ...
        0.0694, 0.4, 1, 0, 0, 0)
```

And we obtain the value for the `call` option as

Stockprice =

50.0000	55.5583	61.7344	68.5971	76.2228	84.6961	94.1113
0	44.9978	50.0000	55.5583	61.7344	68.5971	76.2228
0	0	40.4961	44.9978	50.0000	55.5583	61.7344
0	0	0	36.4447	40.4961	44.9978	50.0000
0	0	0	0	32.7986	36.4447	40.4961
0	0	0	0	0	29.5173	32.7986
0	0	0	0	0	0	26.5643

Optionprice =

5.9095	9.1275	13.6517	19.6280	26.9124	35.0421	44.1113
0	2.6881	4.6100	7.7067	12.4241	18.9432	26.2228
0	0	0.7521	1.4948	2.9708	5.9043	11.7344
0	0	0	0	0	0	0
0	0	0	0	0	0	0
0	0	0	0	0	0	0
0	0	0	0	0	0	0

That is, the `call` option is 5.9095.

MATLAB can generate a tree using the function `CRRTree` available in the Financial Derivatives Toolbox. The general form is

```
CRRTree = crrtree(StockSpec, RateSpec, TimeSpec)
```

We now describe each of these arguments.

1. The argument `StockSpec` creates the structure of the stock. For our purpose, its form can be simply

```
Stock_Specification = stockspec(Sigma, AssetPrice)
```

Where `Sigma` is the volatility and `AssetPrice` is the initial stock price. For our Example 12.7,

```
>> Stock_Specification = stockspec( 0.4, 50)
```

2. `RateSpec` can be generated with `intenvset` which sets the properties of the interest-rate structure. For the data of Example 12.7 we set

```
>> RateSpec = intenvset('Compounding', -1, 'Rates', 0.1, ...
      'StartDates', '01-Jan-2020', 'EndDates', ...
      '31-May-2020', 'Basis', 3)
```

which indicates a 10% risk-free interest rate taking a year of 365 days
and starting on Jan. 1st, 2020, and ending on May 31st, 2020.

3. Finally, the time specification is created with **crrtimespec** whose structure is

```
TimeSpecification = crrtimespec(ValuationDate, ...
      Maturity, NumPeriods)
```

where ValuationDate is the starting date, Maturity is the ending date,
and NumPeriods is the number of periods where we are going to evaluate
the tree.

For our example, the number of periods is 5 monthly periods and

```
>> ValuationDate = '1-Jan-2020'; Maturity = '1-June-2023';
```

Thus,

```
>> TimeSpec = crrtimespec(ValuationDate, Maturity, 5)
```

Running each of these instructions will produce:

```
>> Stock_Specification = stockspec( 0.4, 50)
   Stock_Specification =
   FinObj: 'StockSpec'
   Sigma: 0.4000
   AssetPrice: 50
   DividendType: [ ]
   DividendAmounts: 0
   ExDividendDates: [ ]

>> RateSpec = intenvset('Compounding', -1, 'Rates', 0.1, ...
         'StartDates','01-Jan2020', 'EndDates', ...
            '31-May-2020', 'Basis', 3)

   RateSpec =
   FinObj: 'RateSpec'
   Compounding: 365
```

```
Disc: 0.9597
Rates: 0.1000
EndTimes: 150
StartTimes: 0
EndDates: 731732
StartDates: 731582
ValuationDate: 731582
Basis: 3
EndMonthRule: 1
```

```
>> TimeSpec = crrtimespec('1-Jan-2023', '1-June-2023', 5)
TimeSpec =
FinObj: 'BinTimeSpec'
ValuationDate: 731582
Maturity: 731733
NumPeriods: 6
Basis: 0
EndMonthRule: 1
tObs: [0 0.0691 0.1383 0.2074 0.2766 0.3457 0.4148]
dObs: [731582 731607 731632 731657 731682 731707 731733]
```

Note that dates have been converted to serial date numbers format (see Ch. 3 for date formats). We are now ready to generate the data for the tree with

```
>> CRRTree = crrtree(Stock_Specification, RateSpec, TimeSpec)
```

which produces

```
CRRTree =
FinObj: 'BinStockTree'
Method: 'CRR' StockSpec: [1x1 struct]
TimeSpec: [1x1 struct]
RateSpec: [1x1 struct]
tObs: [0 0.0691 0.1383 0.2074 0.2766 0.3457 0.4148]
dObs: [731582 731607 731632 731657 731682 731707 731733]
STree: 1x7 cell
UpProbs: [0.5067 0.5067 0.5067 0.5067 0.5067 0.5067]
```

In this output data we see the information of the tree. The tObs variable stores the observation points, dObs contains the dates of the observations in serial date format, and UpProbs contains the probability of an up movement. Now, we can plot the tree with the tree viewer instruction

```
>> treeviewer(CRRTree)
```

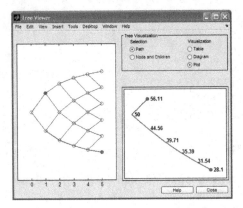

FIGURE 12.11: Binomial tree.

which generates the tree shown in Figure 12.11. In this plot we have selected the nodes for the down movement of the tree. We can compare this result with those obtained from the function `binprice` and we see that they agree very well. For example, the same end value for the tree which is 26.66 is 26.5646 in `binprice`.

We can also create an instrument portfolio by using

```
>> InstSet = instadd('OptStock', {'call'; 'put'}, ...
        50, ValuationDate, Maturity)
```

The names of the variables we wish to display are

```
>> Names = {'Call Option'; 'Put Option'}
```

So now we update the instrument portfolio with

```
>> InstSet = instsetfield(InstSet, 'Index', 1:2, 'FieldName', ...
        'Name', 'Data', Names);
```

And create the tree with

```
>> [Price, PTree] = crrprice(CRRTree, InstSet)
```

`Price` is a vector that contains the `call` option and the `put` option values, respectively. Finally, we display the tree with

```
>> treeviewer( PTree , Names )
```

The tree is displayed in Figure 12.12 for the `put` option. This tree looks like the one shown in Figure 12.11, but it has a pull down menu to display the

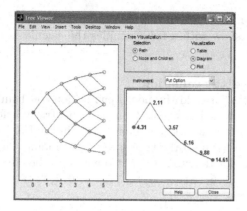

FIGURE 12.12: Binomial tree.

call and put options. We can compare with the results obtained using the function binprice and there we can see that both results agree very well.

12.5 The Greek Letters

MATLAB allows us to calculate the so-called Greek letters which are useful to traders so they manage the risks to make them acceptable. MATLAB can evaluate the delta, gamma, lambda, rho, and theta. It can also evaluate the vega, which is not a Greek letter but falls in the set of the same functions. The instructions to obtain them are the following:

```
[CallDelta, PutDelta] = blsdelta(Price, Strike, Rate, ...
        Time, Volatility, Yield)

Gamma = blsgamma(Price, Strike, Rate, Time, Volatility, Yield)

[CallEl, PutEl] = blslambda(Price, Strike, Rate, Time, ...
        Volatility, Yield)

[CallRho, PutRho] = blsrho(Price, Strike, Rate, Time, ...
        Volatility,Yield)

[CallTheta, PutTheta] = blstheta(Price, Strike, Rate, Time, ...
        Volatility, Yield)

Vega = blsvega(Price, Strike, Rate, Time, Volatility, Yield)
```

The parameters in these instructions are the same as in the case of the Black-Scholes case. We show with a few examples the calculation of some of these Greeks.

Example 12.8 Delta evaluation
A bank in the USA has sold six month put options on 1 million dollars with a strike price of 1.6 and it wishes to make a neutral portfolio. The current exchange rate is 1.62, the risk free interest rate in the UK is 13% per annum, the risk free interest rate in the USA is 10% per annum, and the volatility is 15% [3]. The delta for the call and put options can be obtained using Price = 1.62, Strike = 1.6, Rate = 0.10, Time = 0.5, Volatility = 0.15, and Yield = 13%. Thus,

>> [CallDelta, PutDelta]=blsdelta(1.62,1.6,0.1,0.5,0.15,0.13)

which gives

    CallDelta =
            0.4793

    PutDelta =
            -0.4578

The delta for the put option is -0.4578. It means that when the exchange rate increases by S, the price of the put goes down by 45.78% of S. Thus, the bank has to add a short position of 457800 to the option position to make the delta neutral.

Example 12.9 Theta, Gamma, Vega, Rho, and Lambda evaluation
Consider a 4 month put option on a stock index. The current value for the index is 305, the strike price is 300, the yield is 3% per annum, the interest rate is 8% per annum, and the volatility is 25% per annum[4]. That is, Price = 305, Strike = 300, Rate = 0.08, Time = 4/12, Volatility = 0.25, Yield = 3%. Then, the Theta, Gamma, Vega, Rho, and Lambda can be evaluated with

>> [CallTheta, PutTheta]=blstheta(305,300,0.08,4/12,0.25,0.03)

    CallTheta =
            -32.4623

---

[3]This example is taken from Example 14.1 in [2] with permission of the author.
[4]This example is taken from Examples 14.3-14.5, 14.7 in [2], with permission of the author.

```
PutTheta =
          -18.1528

>> Gamma = blsgamma(305, 300, 0.08, 4/12, 0.25, 0.03)

   Gamma =
          0.0086

>> Vega = blsvega(305, 300, 0.08, 4/12, 0.25, 0.03)

   Vega =
         66.4479

>> [CallRho, PutRho] = blsrho(305, 300, 0.08, 4/12, 0.25, 0.03)

   CallRho =
           54.7893
   PutRho =
          -42.5792

>> [Call_Lambda, Put_Lambda] = blslambda(305, 300, 0.08, ...
        4/12, 0.25, 0.03)

   Call_Lambda =
              8.3156

   Put_Lambda =
              -9.1310
```

This means that the **put** option value decreases by **Theta**/365 per day, that is, $42.5792/365 = 0.0497$ per day. Also, an increase of 1 in the index increases the **delta** of the option by **Gamma** = 0.0086, approximately. In addition, an increase of 1% in the volatility increases the value of the option by (1%***Vega** =) 66.4479, approximately. The value of **Rho** implies that a 1% change in the interest rate decreases the value of the **put** option by $(0.01*42.6 =)$ 0.426. Finally, the **lambda** value measures the percent change in the **put** option price per one percent change in the asset price.

Example 12.10 Plots of the Greeks

We can make plots of the Greeks against the several variables that are involved in their computation. For example, if we wish to plot the **Rho** against the interest rate and the asset price we can do this with the following script:

Plot of Rho for a Put Option

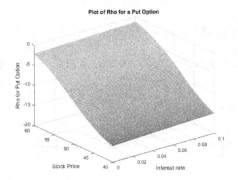

FIGURE 12.13: Plot of Rho for a put option vs. interest rate and stock price.

Plot of Delta for a Put Option

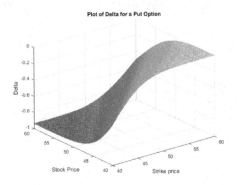

FIGURE 12.14: Plot of delta for a put option vs. strike price and stock price.

```
% plot of Rho
r = 0: 0.001: 0.1;
S = 40:.2:60;
for i = 1:length(r)
    for j = 1:length(S)
        [CallRho(i,j), PutRho(i,j)] = blsrho(S(i), 50, ...
        r(j), 4/12, 0.25, 0.03);
    end
end
mesh(r, S, PutRho)
title('Plot of Rho for a Put Option')
xlabel('Interest rate'); ylabel('Stock Price'); zlabel('Rho for
Put Option')
```

The resulting plot is shown in Figure 12.13.

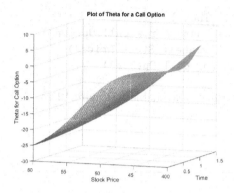

FIGURE 12.15: Plot of Theta for a call option vs. interest rate and stock price.

For a plot of delta vs. the stock price and the strike price we can use the following script to obtain Figure 12.14.

```
% plot of delta
close all; clear all; clc
K = 40: 0.2: 60;
S = 40: 0.2: 60;
for i = 1:length(K)
    for j = 1:length(S)
        [Call_Delta(i,j), Put_Delta(i,j)] = ...
            blsdelta(S(j), K(i), 0.10, 0.5, 0.15, 0.13);
    end
end
mesh(K, S, Put_Delta)
title('Plot of delta for a Put Option')
xlabel('Strike price'); ylabel('Stock Price'); zlabel('delta')
```

Now if we wish to plot Theta for a call option versus stock price and time we use:

```
clear; close all; clc
T = 0.00001: 0.01: 1.01;
S = 40: 0.2: 60;
for i = 1: length(T)
    for j = 1: length(S)
        [CallTheta(i,j), PutTheta(i,j)] = blstheta(S(j), ...
            50, T(i), 0.5, 0.15, 0.13);
    end
end
```

```
mesh(T, S, CallTheta)
title('Plot of Theta for a Call Option')
xlabel('Time'); ylabel('Stock Price')
zlabel('Theta for Call Option')
```

The resulting mesh plot is available in Figure 12.15.

12.6   Conclusions

We have presented techniques to solve financial problems using MATLAB. MATLAB's potential is best appreciated in computing intensive techniques such as the Black-Scholes computations, sensitivity calculation, and plotting. The examples provided here are merely representative of the finance functions that MATLAB can compute and, as always, it is up to the reader to take full advantage of MATLAB. Further information on the functions available in both the Financial toolbox and the Financial Derivatives toolbox is available in the User's Guide available at The Mathworks, Inc. website www.mathworks.com.

12.7   References

[1] F. Black and M. Scholes, The Pricing of Options on Corporate Liabilities, J. Political Economic, May-June 1973, pp. 637-654.

[2] J. Hull, Options, Futures and Other Derivatives, Prentice-Hall Book Co., 2017.

[3] J.C. Cox, S. A. Ross, and M. Rubinstein, Option Pricing: A Simplified Approach, J. Financial Economics, No. 7, Oct. 1979, pp.229-264.

[4] M. Hoyle, Pricing American Options, available at: https://la.mathworks.com/matlabcentral/fileexchange/16476-pricing-american-options, consulted on May 22, 2018.

[5] P. Brandimarte, Numerical Methods in Finance: A MATLAB-Based Introduction, J. Wiley & Sons, New York, 2006.

[6] F. A. Longstaff and E. S. Schwartz, "Valuing American Options by Simulation: A Simple Least-Squares Approach", May 9, 2001, The Review of Financial Studies, Vol. 14, No. 1, pp. 113-147.

Chapter 13

Image Processing in MATLAB

13.1   Introduction

A black and white image is a two-dimensional representation of three-dimensional information in color. For processing, an image is partitioned into a number of elements called pixels. This word comes from the combination of the English words PICture ELement. The standard size of images in modern devices is in the range of megapixels. In Figure 13.1, it is shown how an image is represented. From this figure it can be seen that there are $8 \times 8 = 64$ pixels. A natural way of representing an image is then by a matrix where each pixel position is associated with the values n and m of the matrix elements. The value of the element m and n of the matrix is then the value of gray level of the corresponding pixel.

The way to number the pixels is exactly equal to the representation of matrices as shown in Figure 13.2. Thus, the 0,0 pixel is a white pixel and pixel 3,5 is a black pixel. Each pixel has an intensity level. For monochrome images (black and white) the levels of intensity are gray levels. Depending on how many bits are handled for this representation, it will be the number of gray levels that can be represented. The image of Figure 13.2 is a chessboard that only has two levels and therefore can be represented by a single bit. For example, if that bit is 0 then it is black and if the pixel is 1 it is white, and vice versa. So that images can be represented adequately then at least 256 gray levels are needed which requires that each pixel is represented with 8-bit words. In the case of color images, a word of 8 bits is required to represent

FIGURE 13.1: Pixels of an image.

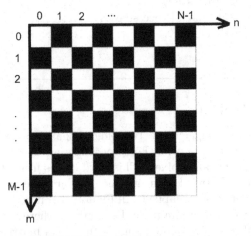

FIGURE 13.2: Ordering of pixels.

each of the three color components, i.e., 24 bits are required for each color pixel[1].

MATLAB allows the processing of images by using the Image Processing package. Images can be in different formats, such as bmp, eps, tif, pcx, gif, png, etc. and even in a free format. To test our instructions in MATLAB we will use an image known as Lena that researchers use as a reference. This image of Lena is shown in Figure 13.3. The original image is chromatic, in a bmp format and has a size of 512×512.

---

[1]Image files that are discussed in this chapter can be found in the web page of the book.

FIGURE 13.3: Lena image. This image is 512×512.

13.2 Reading and Writing Images

The instruction to read images is `imread` ('`image`'), where `image` specifies the location and file name. If the image is located in the Documents folder (for Windows) and the file is `lena.bmp`, then:

```
f = imread('C:\Documents\lena.bpm');
```

This puts the image into variable `f` and leaves it ready for processing by MATLAB. The image can be displayed with the instruction `imshow` which has the format:

```
imshow(f)
```

The image of Lena is in an RGB color format. The instruction `rgb2gray` is used to convert it to gray levels

```
lena_gray = rg2gray(f);
```

and now the image is in gray levels. The instruction `indexsize` is used to view the size of an image. For two images then:

```
>> size(f)
   ans =
      512 512 3
```

FIGURE 13.4: Part of the image corresponding to `imshow(lena_gray (210: 400, 210: 500))`.

```
>> size(lena_gray)
   ans =
      512 512
```

It can be seen that since the image `f` is chromatic it is represented by three matrices of size 512×512 while `lena_gray` is only a single 512×512 matrix.

Since images are represented as matrices, matrix operations can be used for images. For example, if the following operation is performed:

```
imshow( lena_gray(210:400, 210:500) )
```

we obtain Figure 13.4 showing the part corresponding to the image. In the same way, the instruction `plot` can be used on a row or a column to display the intensity levels or gray levels. For example

```
plot(lena_gray(131, :))
```

produces the graph of gray levels of the row 131 shown in Figure 13.5.

13.3   Resolution of the Images

Images can have a limited dynamic range as in the case of the image shown in Figure 13.6, which is an image in a jpg format. After reading it in MATLAB using

```
X_rays = imread('FigureX_rays.jpg');
```

the histogram can be obtained with the instruction `hist(X_rays, N)` where

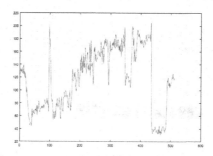

FIGURE 13.5: Gray levels for row 131 for the image lena_gray.

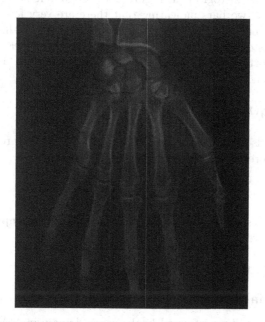

FIGURE 13.6: X-ray image with a limited dynamic range.

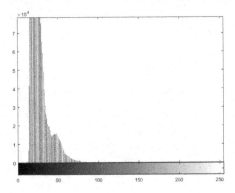

FIGURE 13.7: Histogram of the image of Figure 13.6.

N is the number of bars in the histogram and X_rays is the image. In this case N = 256 and:

```
imhist('X_rays', N);
```

that produces the histogram in Figure 13.7. Clearly it is seen that the histogram has only two bars indicating that there are very few gray levels corresponding to very dark gray levels. One way to correct this problem is to equalize the histogram to reduce the number of gray levels of the first bar and raise proportionally the others that are imperceptible. The equalization is obtained with:

```
g = histeq(X_rays);
```

and the result is shown in Figure 13.8. It is seen that there are more gray levels than before. The equalized image is displayed with

```
imshow(g);
```

and the result is shown in Figure 13.9. We readily see an improvement in the image because we now have more gray levels in it.

13.4 Spatial Filtering

Image filtering can be performed in the space domain in MATLAB with the instruction imfilter whose format is

```
image_b = imfilter(image_to_process, w, options);
```

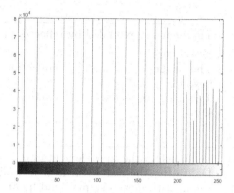

FIGURE 13.8: Equalized histogram.

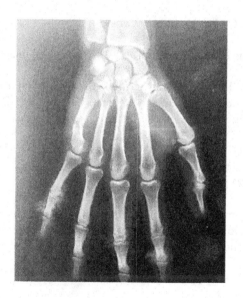

FIGURE 13.9: Equalized image.

TABLE 13.1:   IMFilter options.

Options Method	Descriptions
'Symmetric'	The image is extended by reflecting it beyond its borders.
'Replicate'	The image is extended by replicating its values beyond its borders.
'Circular'	The image extends like if it were a one-dimension signal.
Direction	
'Pre'	Do it before the first element of each dimension.
'Post'	Do it after the last element of each dimension.
'Both'	Using both methods 'pre' and 'post'.

where `image_b` is the target image, `image_to_process` is the image that is to be filtered, `w` is the mask for filtering and `option` is one of those listed in Table 13.1. The filter type is obtained using the instruction `fspecial` that has the format

```
w = fspecial('type', parameters)
```

where 'type' is one of the filters in Table 13.2 (only the most used are shown).

Example 13.1 Filtering using a mask of a disk with a radius 3.
Consider the monochromatic image of Lena. Using a disc filter with radius = 9 and filtering using `imfilter`, the instructions are:

```
w = fspecial('disk', 9);
lena2 = imfilter(lena_gray, w);
imshow(lena2);
```

The result is the image shown in Figure 13.10 that is blurred. Clearly, the filter applied is a low pass filter where fast changes in contrast correspond to high frequencies and they have been attenuated because of the low pass filter.

TABLE 13.2: Types of filters. The format is `fspecial(Parameters)`.

Type	Parameters	Syntax
'average'	'average', [r, c]	Rectangular average filter size r×c. By default, the size is 3×3.
'disk'	'disk', r	Average circular filter of radius r. The default is r = 5.
'gaussian'	'gaussian',[r,c],sig	Gaussian filter of size r with a standard deviation sig. The default values are 3×3 and 0.5.
'laplacian'	'laplacian', alpha	3×3 Laplacian filter with an alpha between 0 and 1. The default value of alpha is 0.5.
'log'	'log', [r, c], sig	Laplacian of a Gaussian (log). The default values are 5×5 and 0.5.
'motion'	'motion', len, theta	The result approximates a filter in movement by len pixels when it is convolved with an image. The default values are len = 9 and theta = 0.

13.5 The Discrete Fourier Transform

The Discrete Fourier Transform is a widely used tool in image filtering. It is evaluated by the algorithm of the Fast Fourier Transform. The instruction is `fft2` which stands for Fast Fourier Transform in 2 dimensions. Consider the image shown in Figure 13.11, which is a white square in a black background. To obtain its Fourier Transform, first the image is read into the workspace of MATLAB with `imread`, then the Discrete Fourier Transform is applied using the two-dimensional `fft2`, then it is shifted so that the center of the image is the origin by using `fftshift` and finally it is displayed with `imshow`. To calculate and display the magnitude of the Fourier Transform:

```
subplot(2, 2, 1)
F = fft2(square)
S = abs(F);
imshow(S, [ ])
title('Fourier Transform')
```

It is observed that most information occurs at the corners of the figure. The

FIGURE 13.10: Filtered image of a disc with radius 9.

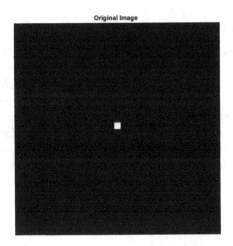

FIGURE 13.11: A square to obtain the Discrete Fourier Transform.

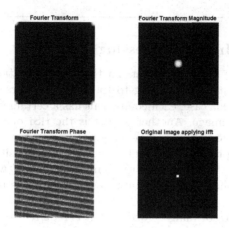

FIGURE 13.12: Obtaining the Fourier Transform in two dimensions.

instruction `fftshift(S)` is used to center the information. Then the instructions are:

```
subplot(2, 2, 2)
Fc = fftshift(F);
imshow(abs(Fc), [ ])
title('Fourier Transform magnitude')
```

The phase can be plotted with

```
subplot(2, 2, 3)
imshow(angle(Fc), [ ])
title('Fourier Transform phase')
```

Finally, the inverse transform is calculated with

```
subplot(2, 2, 4)
imshow(ifft2(F))
title('Original image.')
```

and we obtain the original figure. The results of these actions are shown in Figure 13.12.

13.6   Color Image Processing

Color images have different formats for their representation. The best known is the RGB format corresponding to the primary colors red (R), green (G) and blue (B). In this case, each component has a certain intensity to produce the colors in the image. Another format is the HSI one consisting of three components that are the hue (H) corresponding to the tone, the saturation (S) corresponding to the amount of color, and the intensity (I) corresponding to how intense is the color. Another format is the HSV where V is for value (V). There are other color formats whose description is beyond the scope of this chapter.

To perform a color image processing, it is required to do a decomposition into the three basic components. For example, for the image of Lena, the three RGB components are obtained with

```
figure
subplot(2, 2, 1)
imshow(lena)
title('Original image')
subplot(2, 2, 2)
lena_red = lena(:, :, 1);
imshow(lena_red)
title('Red component')
subplot(2, 2, 3)
lena_green = lena(:, :, 2);
imshow(lena_green)
title('Green component')
subplot(2, 2, 4)
lena_blue = lena(:, :, 3);
imshow(lena_blue)
title('Blue component')
```

With these instructions Figure 13.13 is produced. The different intensities of each of the color components in the image of Lena can now be appreciated.

The instruction rgh2hsv is used to convert an RGB image to an HSV image. In particular, for the Lena in RGB it is used

```
lena_HSV = rgb2hsv(lena);
```

As in the RGB format, the HSV format image and its components can be plotted, as shown in Figure 13.14. Note that the image in the HSV format has different features compared to those of the RGB format.

FIGURE 13.13: Lena image with its RGB components.

FIGURE 13.14: Lena in the HSV format and its components.

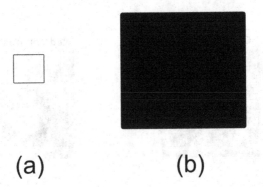

(a)　　　　　　　　　(b)

FIGURE 13.15: Structuring element and image to be processed.

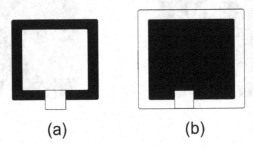

(a)　　　　　　　　　(b)

FIGURE 13.16: Result from an (a) erosion, (b) dilation. The original image is the black square. The structuring element is the small square.

13.7 Morphological Image Processing

Morphological image processing is based on mathematical morphology. Thus, it is based in set theory. It operates mainly on binary images. Morphological image processing has the basic operations of erosion and dilation. These operations require using a structuring element B on an image A. Let us assume that the structuring element is a disk of radius r as shown in Figure 13.15a and the image to process is in Figure 13.15b.

An erosion consists in shifting the structuring element center around the perimeter of the image at the inner part by removing from the figure everything over where the structuring element overlaps the image. This action is shown in Figure 13.16a. The result is that the image decreases in size, and if there were any imperfections in the sizes of the square, they would be smoothed.

Dilation is an opposite action and in this case the structuring element is shifted at the outside as shown in Figure 13.16b, increasing the size of the image and also changing the imperfections, if any.

FIGURE 13.17: Test image for the four morphological operations.

FIGURE 13.18: Results of the morphological operations: erosion, dilation, opening and closing.

From these basic operations, the combination of an erosion followed by a dilation is known as an opening and is denoted by A⊙B. The combination of a dilation followed by an erosion is denoted by AB and is called a closure. The image processing package has included among its functions these four morphological operations. As a test let us use the image of Figure 13.17.

This figure shows some black letters on a white background with salt and pepper noise (black and white dots) and black spots. With the morphological filtering the results are shown in Figure 13.18. In some cases, they may be repeated for further processing operations.

The instructions for performing morphological filtering are:

```
Aerosion = imerode(image, structuring_element);
ADilation = imdilate(image, structuring_element);
Aopening = imopen(image, structuring_element);
Aclosing = imclose(image, structuring_element);
```

Results are displayed with:

```
imshow(Aerosion)
imshow(Adilation)
imshow(Aopening)
imshow(Aclosing)
```

Two other important concepts in morphological processing are thinning and

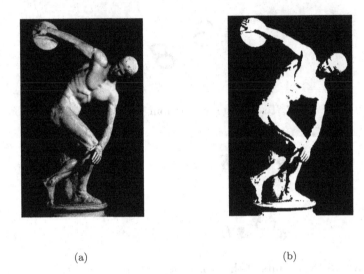

(a)   (b)

FIGURE 13.19: Discus Thrower of Myron. a) Original image, b) Binarized image.

skeletonization. This technique is useful for finding basic structures. To do this the instruction used for skeletonization is

```
A = bwmorph(image, 'skel', n);
```

And for thinning we use:

```
A = bwmorph(image, 'thin', n);
```

where n is an integer that specifies how many times the operation is repeated. If n = Inf is used, the operation is repeated until the result no longer changes. Consider the image of the Discus Thrower of Myron in Figure 13.19a.

First, this figure is binarized with:

```
D = imread('disk_thrower.eps'); % We read the image.
BW = im2bw(D, 0.3); % We binarize with a level 0.3.
imshow(BW) % We display the image.
```

The resulting image is shown in Figure 13.19b. Now the thinning is made three times with

```
A = bwmorph(BW, 'thin', 3);
```

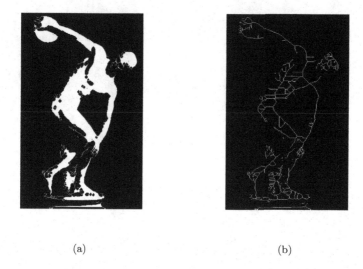

(a)   (b)

FIGURE 13.20: Thinning of discus thrower. a) Three times, b) Infinite number of times.

This produces Figure 13.20a. Now, it is repeated over the original image but with n = Inf obtaining Figure 13.20b.

13.8   Concluding Remarks

This chapter provides an introduction to the MATLAB image processing package. With a set of examples, an idea of how different techniques can be applied to image processing in an expeditious and simple manner was given.

Index

tone, 348
toolbar, 214
Toolbar Editor, 214
Tools, 214
trace, 29, 99
traces, 36
Transfer Function, 230
transpose, 99
tree construction, 319
treeviewer, 329
triangle, 34
trigonometric functions, 5
type, 9

uint16, 10
uint32, 10
uint64, 10
uint8, 10
unit bond price, 319
upper, 12, 13
upper triangular, 109
upward, 34

variable, 61
variables, 7
vector, 8, 99
vector product, 105
Vega, 319
vega, 331
viewpoint, 56
visualization, 23
volatility, 319, 320

waterfall, 51
while, 144
white, 35
whitebg, 35
who, 7
width, 33
Workspace, 30
Workspace window, 3

x-axis, 27
xlabel, 27, 29, 66

y-axis, 27

yield, 311, 317
ylabel, 27, 29, 66

zero curve, 319
zeros, 74, 95
Zoom, 231

Printed in the United States
by Baker & Taylor Publisher Services